Contract No. EP-D-06-003
Work Assignment No. 4-84
RTI Project Number
0212979.001.002

Regulatory Impact Analysis (RIA) for Proposed Residential Wood Heaters NSPS Revision

Final Report

January 2014

Prepared for

Larry Sorrels
U.S. Environmental Protection Agency
Office of Air Quality Planning and Standards (OAQPS)
Health and Environmental Impacts Division (HEID)
Air Economics Group (AEG)
(MD-C439-02)
Research Triangle Park, NC 27711

Prepared by

Jeffrey Petrusa
Stephanie Norris
Brooks Depro
RTI International
3040 Cornwallis Road
Research Triangle Park, NC 27709

CONTENTS

LIST OF FIGURES

LIST OF TABLES

SECTION 1
EXECUTIVE SUMMARY

The U.S. Environmental Protection Agency (EPA) is proposing to revise new source performance standards (NSPS) for residential wood stoves, and to issue NSPS for pellet stoves, furnaces, hydronic heaters, and masonry heaters. The EPA is proposing this revision under the authority of section 111 of the Clean Air Act (CAA), "Standards of Performance for New Stationary Sources," under which the EPA establishes federal standards of performance for new sources within source categories which cause or contribute significantly to air pollution, which may reasonably be anticipated to endanger public health or welfare. We are proposing to amend 40 CFR part 60, subpart AAA, Standards of Performance for New Residential Wood Heaters. The current regulation (subpart AAA) applies to affected facilities manufactured since 1988. Except as discussed in this proposal, the current requirements would remain in effect for the heaters/stoves and model lines manufactured before this proposal. We also propose to broaden the applicability of the wood heaters regulation beyond adjustable burn rate heaters (stoves, the focus of the original regulation) to specifically include single burn rate heaters, hydronic heaters, and pellet stoves. Heaters/stoves and model lines manufactured after the compliance dates would be required to meet particulate matter (PM) standards. Compliance upon the effective date of the final rule is the intention in section 111 of the CAA. Revision of the current residential wood heaters NSPS is necessary to capture the improvements in performance of such units and to include additional wood-burning residential heating devices. The proposed changes are expected to achieve several objectives, including the application of updated emission limits reflecting the best industry emission reduction systems; elimination of exemptions over a broad suite of residential wood combustion devices; the strengthening of test methods as appropriate; and the streamlining of the certification process. This proposal does not include any requirements for heaters solely fired by gas, oil or coal. In addition, it does not include any requirements associated with wood heaters or other wood-burning appliances that are already in use. The EPA continues to encourage state, local, tribal, and consumer efforts to change out (replace) older heaters with newer, cleaner, more efficient heaters, but that is not part of this Federal rulemaking. These proposed revisions help address the health impacts of particle pollution, of which wood smoke is a contributing factor in many areas. Particulate pollution from wood heaters is a significant national air pollution problem and human health issue. Health benefits associated with these proposed regulations are valued to be much greater than the cost to manufacture cleaner, lower emitting appliances. These proposed regulations would also significantly reduce emissions of many other pollutants from these appliances, including carbon monoxide, volatile organic compounds, hazardous air pollutants and climate-forcing emissions.

Emissions from wood stoves occur near ground level in residential communities across the country, and setting these new requirements for cleaner stoves into the future will result in substantial reductions in exposure and improved public health.

Wood smoke contains a mixture of fine particles and toxic air pollutants (e.g., benzene and formaldehyde) that can cause burning eyes, runny nose, and bronchitis. Exposure to fine particles has been associated with a range of health effects, including aggravation of heart or respiratory problems, changes in lung function and increased respiratory symptoms, as well as premature death. Populations that are at greater risk for experiencing health effects related to fine particle exposures include older adults, children and individuals with pre-existing heart or lung disease. Each year smoke from wood heaters and fireplaces contributes hundreds of thousands of tons of fine particles throughout the country—mostly during the winter months. For more information on the health impacts from exposure to fine particles, please refer to Section 7 of this RIA. Nationally, residential wood combustion accounts for 44% of total stationary and mobile polycyclic organic matter (POM) emissions, nearly 25 percent of all area source air toxics cancer risks and 15 percent of noncancer respiratory effects. Residential wood smoke causes many counties in the U.S. to either exceed the EPA's health-based national ambient air quality standards (NAAQS) for fine particles or places them on the cusp of exceeding those standards. For example, in places such as Keene, New Hampshire; Sacramento, California; Tacoma, Washington; and Fairbanks, Alaska; wood combustion can contribute over 50 percent of daily wintertime fine particle emissions. The concerns are heightened because wood stoves, hydronic heaters, and other heaters are often used around the clock in many residential areas. To the degree that older, dirtier, less efficient wood heaters are replaced by newer heaters that meet the requirements of this rule, or better, the emissions would be reduced, the efficiencies would be increased, and fewer health impacts should occur.

This is an economically significant rule as defined by Executive Order 12866 and Executive Order 13563. Therefore, EPA is required to develop a regulatory impact analysis (RIA) as part of the regulatory process. The RIA includes an economic impact analysis (EIA), a small entity impacts analysis, an engineering cost analysis, and a benefits analysis along with documentation for the methods and results.

We provide annualized average results for the time frame from 2014 to 2022 inclusive for two options: the Proposed option and an Alternative option. While the Proposed Option represents EPA's preferred option, the Alternative Option is still under consideration. These options are described in detail in the preamble for the proposal and in Section 2 of this RIA and summarized in Section 4. The options vary in part by their respective dates of implementation, all of which are captured by the range of dates included in the analyses. We estimate the impacts for the time frame from 2014 to 2022 in order to provide an average of annualized results for these options from the time of rule promulgation in 2014 to the time of full implementation of both options, which occurs by 2022. Because the potential environmental impacts can occur for 40 years or more, which is the typical useful life for wood heater appliances, the impacts for 40 years are also shown in the appendix within Section 9 of this RIA. The variability of annual impacts for each option provides an appropriate rationale for presenting impacts averaged over this time frame. All results in this RIA are presented in 2010 dollars. Estimates of benefits and costs are discounted to the analysis year using both a 7% and 3% discount rates following Circular A-4, "Regulatory Analysis," which provides guidance to Federal agencies on the development of regulatory analyses required by Executive Order 12866.[1]

In addition, this proposal cannot be certified as not having a significant economic impact on a substantial number of small entities (SISNOSE) according to the provisions of the Small Business Regulatory Enforcement Fairness Act (SBREFA). Therefore, small entity impacts analysis presented in Section 6 is an Initial Regulatory Flexibility Analysis (IRFA). Section 6 also contains a summary of the proceedings and conclusions of a panel called to find ways to mitigate small entity impacts associated with this rule under the authority of the SBREFA.

1.1 Analysis Summary

The key results of the RIA are as follows:

- **Engineering Cost Analysis:** EPA estimates the revised NSPS's total annualized cost to affected manufacturers on average in the 2014–2022 time frame will be $15.7 million ($2010) for the Proposed option and $28.3 million (2010$) for the Alternative option, respectively, with the total annualized cost estimate at a 7% discount rate. At a 3% discount rate, the total annualized cost will be $14.8 million for the Proposed option and $26.9 million for the Alternative option.

- **Economic Impact Analysis:** The metric for economic impacts for industries affected by this Proposed option are industry-level average annualized compliance costs to receipts (or sales) ratios. This metric is calculated as an average in the 2014–2022 time frame, and the estimates ranged from 4.3% for industries that produce wood stoves to as much as 6.4% for single burn rate stoves. For the Alternative option, the range is between 4.0% for forced air furnaces and 10.7% for single burn rate stoves. These results approximate the maximum price increase needed for a producer to fully recover the annual compliance costs and, therefore, do not presume any pass through

---

[1] Circular A-4 is available at: http://www.whitehouse.gov/omb/circulars_a004_a-4

of impacts to consumers. With pass through to consumers, these impact estimates will decline proportionately to the degree of pass through.

- **Social Cost Analysis:** For this RIA, the Agency assumes that the social cost is equal to the annualized cost to manufacturers. Therefore, the estimated average annual social cost of the Proposed and Alternative options in the 2014-2022 timeframe are expected to be $15.7 million and $14.8 million respectively when discounted at 7% , and $28.3 million and $26.9 million when discounted at 3%. See Section 5 of this RIA for more detail on the estimated social cost.

- **Small Entity Analyses:** EPA performed a screening analysis for impacts on small entities by comparing compliance costs to sales/revenues (e.g., sales and revenue tests). EPA's analysis showed the tests were higher than 1% for small entities included in the screening analysis; the 1% test estimate is often an indicator for significant impacts to small firms. For these industries, almost all (more than 90%) affected entities are small firms. We concluded that we could not certify that there would not be a significant economic impact on a substantial number of small entities (SISNOSE) for either option considered in this RIA. Pursuant to section 603 of the RFA, EPA prepared an initial regulatory flexibility analysis (IRFA) for the proposed rule and convened a Small Business Advocacy Review Panel to obtain advice and recommendations of representatives of the regulated small entities. A detailed discussion of the Panel's advice and recommendations is found in the final Panel Report (Docket ID No. EPA-HQ-OAR-2009-0734. A summary of the Panel's recommendations is also presented in the preamble to the proposal. In the proposal, EPA included provisions consistent with several of the Panel's recommendations.

- **Benefits Analysis**:

  - Monetized benefits in this RIA include those from reducing particulate matter (PM). These benefits reflect reductions of nearly 4,800 tons annually of fine particulate matter ($PM_{2.5}$) on average during the 2014–2022 time frame under each of the two options. All monetized benefits reported reflect improvements in ambient $PM_{2.5}$ concentrations due to emission reductions of direct $PM_{2.5}$. As a result, the monetized benefits likely underestimate the total benefits, however, the extent of the underestimate is unclear. Monetized benefits reflect those associated with reductions in premature mortality due to lower ambient $PM_{2.5}$ concentrations resulting from implementation of the NSPS. Other benefits categories from $PM_{2.5}$ reductions, such as changes in visibility, are assessed qualitatively in this analysis.

  - Using a 3% discount rate, we estimated the total monetized benefits of the Proposed and Alternative options to be $1.8 billion to $4.2 billion and $1.9 billion to $4.2 billion, respectively based on estimates by Krewski and Lepeule, in the 2014–2022 time frame. Using a 7% discount rate, we estimate the total monetized benefits of the Proposed option to be $1.7 billion to $3.8 billion and $1.7 billion to $3.8 billion for the Alternative option in the 2014–2022 time frame. The benefits are almost identical for both options analyzed. Using alternative relationships between $PM_{2.5}$ and premature mortality supplied by experts, higher and lower benefits estimates are plausible, but most of the expert-based estimates fall between these estimates.

— The benefits from reducing some air pollutants have not been monetized in this analysis due to data and resource constraints, including reducing 33,000 tons of carbon monoxide (CO), over 3,200 tons of volatile organic compounds (VOCs), and undetermined amounts of black carbon and HAP emissions under each option analyzed. Data, resources, and methodological limitations prevented EPA from monetizing the benefits from these important benefit categories. We assessed the benefits of these emission reductions qualitatively in this RIA.

- **Net Benefits:** For the residential wood heater NSPS, the net benefits (benefits minus the costs) are $1.8 billion to $4.1 billion ($2010) at a 3% discount rate and $1.7 billion to $3.7 billion ($2010) at a 7% discount rate for the Proposed option and $1.8 billion to $4.2 billion ($2010) at a 3% discount rate and $1.7 billion to $3.8 billion ($2010) at a 7% discount rate for the Alternative option in the 2014–2022 time frame. All net benefits are in 2010 dollars ($2010).

1.2 Organization of this Report

The remainder of this report supports and details the methodology and the results of the RIA:

- Section 2 describes the proposed regulation.

- Section 3 presents the profile of the affected industries.

- Section 4 describes the baseline emissions and emission reductions for the alternatives analyzed in this proposal.

- Section 5 describes the engineering costs, economic impacts, analyses to comply with Executive Orders, and employment impacts.

- Section 6 describes the small entity impact analyses and the Initial Regulatory Flexibility Analysis (IRFA) prepared by EPA.

- Section 7 presents the benefits estimates.

- Section 8 presents the net benefits (benefits minus costs) for the alternatives analyzed in this proposal.

- Section 9 presents references for the RIA and documentation on the cost analysis and estimates of costs and emission reductions for the proposal under each option beyond 2022.

Table 1-1. Summary of the Monetized Benefits, Social Costs, and Net Benefits for the Proposed Residential Wood Heaters NSPS in the 2014–2022 Time Frame ($2010 millions)[a]

	3% Discount Rate			7% Discount Rate		
Proposed Option						
Total Monetized Benefits[b]	$1,800	To	$4,200	$1,700	to	$3,700
Total Social Costs[c]		$15			$16	
Net Benefits	$1,800	To	$4,100	$1,700	To	$3,700
Nonmonetized Benefits	32,600 tons of CO					
	3,200 tons of VOC					
	Reduced exposure to HAPs, including formaldehyde, benzene, and polycyclic organic matter					
	Reduced Climate effects due to reduced black carbon emissions					
	Ecosystem effects					
	Reduced visibility impairment					
Alternative Option						
Total Monetized Benefits[b]	$1,900	To	$4,200	$1,700	To	$3,800
Total Social Costs[c]		$27			$28	
Net Benefits	$1,800	To	$4,200	$1,700	to	$3,800
Nonmonetized Benefits	32,900 tons of CO					
	3,200 tons of VOC					
	Reduced exposure to HAPs, including formaldehyde, benzene, and polycyclic organic matter					
	Reduced Climate effects due to reduced black carbon emissions					
	Ecosystem effects					
	Reduced visibility impairment					

[a] All estimates reflect average annual estimates for the time frame from 2014 to 2022 inclusive, and are rounded to two significant figures. These results include appliances anticipated to come online and the lowest cost disposal assumption. Total annualized costs are estimated at a 7% and at a 3% interest rate.

[b] The total monetized benefits reflect the human health benefits associated with reducing exposure to $PM_{2.5}$ through reductions of directly emitted $PM_{2.5}$. It is important to note that the monetized benefits include many but not all health effects associated with $PM_{2.5}$ exposure. Benefits are shown as a range from Kreuski et al. (2009) to Lapeule et al. (2012). These models assume that all fine particles, regardless of their chemical composition, are equally potent in causing premature mortality because the scientific evidence is not yet sufficient to allow differentiation of effect estimates by particle type. Because these estimates were generated using benefit-per-ton estimates, we do not break down the total monetized benefits into specific components here. See Figure 7-1 for an illustration of the breakdown, or the RIA for the final Cross-States Air Pollution Rule (EPA, 2011) for more information.

[c] The annualized social costs are $14.8 million for the Proposed option at a 3% discount rate and $26.9 million for the Alternative option when calculated at a 7% interest rate.

SECTION 2

INTRODUCTION

2.1 Background for Proposed Rule

EPA is considering amending the New Source Performance Standard (NSPS) for new residential wood heaters. EPA promulgated the original NSPS for new residential wood heaters including wood stoves in 1988. Based on a review of the NSPS in 2009, EPA noted significant technological improvements that allow emissions from these sources to be better controlled than the current standard. Residential Wood remains one of the five largest categories of PM emissions according to the 2008 National Emissions Inventory.[2] Thus, EPA is proposing to revise the current NSPS standards to improve regulation of wood heaters and broaden the new regulation to cover other residential heating devices. Specifically, EPA is proposing to amend subpart AAA, Standards of Performance for New Residential Wood Heaters. We are also proposing two new subparts to address additional types of wood heating appliances—subpart QQQQ, Standards of Performance for New Residential Hydronic Heaters and Forced-Air Furnaces, and subpart RRRR, Standards of Performance for New Masonry Furnaces. The following sections describe the major proposed provisions of each subpart. Full details on the proposed provisions in each of these subparts can be found in the preamble for this proposal.

2.2 Room Heaters

The current wood heaters regulation (subpart AAA) applies to adjustable burn rate wood heaters/stoves manufactured since 1988. We propose to broaden the applicability of the wood heaters regulation beyond adjustable burn rate wood heaters (the focus of the original regulation) to also specifically include single burn rate wood heaters, and pellet heaters/stoves. These and any other affected appliance would be covered in subpart AAA as a "room heater." We believe this "room heater" categorization better describes the appliances potentially affected under subpart AAA and included in this proposal. Note that this RIA and the proposal use the following terms interchangeably: heaters, stoves, and heaters/stoves. The current emission limits under subpart AAA would remain in effect for the heaters/stoves and model lines manufactured before the effective date of the final rule until their current EPA certification expires (maximum of 5 years) or is revoked. After the certification expires or is revoked, these heaters and other new heaters would have to meet updated emission standards. The proposed subpart AAA exempts new residential hydronic heaters, new residential forced-air furnaces, and new residential masonry heaters because they would be subject to their own subparts. The proposed

---

[2] U.S. EPA, 2008 National Emissions Inventory. Accessed on Sept. 11, 2012.

subpart AAA retains the exemption for fireplaces, strengthens the definition for "cookstoves," and adds definitions for "camp stoves" and "traditional Native American bake ovens" to clarify that they would not be subject to the standard other than a requirement for appropriate labeling for cookstoves and camp stoves. Finally, the proposal clarifies that the emission limits would only apply to wood-burning devices (i.e., not to devices that solely burn fuels other than wood, e.g., gas or oil or coal or other biomass). . In addition, this proposal does not include any requirements associated with wood heaters or other wood-burning appliances that are already in use. The EPA continues to encourage state, local, tribal, and consumer efforts to change out (replace) older heaters with newer, cleaner, more efficient heaters, but that is not part of this Federal rulemaking.

NSPS determinations of the best system of emission reductions (BSER), formerly referred to as best demonstrated technology (BDT), must consider costs (see section II of the preamble for more detail). The fact that this rule applies to consumer products manufactured for sale results in cost considerations that are fundamentally different from most NSPS. Specifically, the cost of potential lost revenues if production and sales had to be suspended while designing and certifying cleaner models would be significant and necessitates reasonable, phased implementation of emission limits. This was true in 1988 and is still true today. Thus, we propose having a transition period so that stoves with currently effective EPA certification can continue to be manufactured until the current certification expires (5 years from date of certification) or is revoked by the Administrator, whichever date is earlier. Renewal of these certifications would not be permitted. That is, in the near term, we are proposing to retain the current 1990 PM emission limits for adjustable burn rate wood heaters and pellet stoves with a current EPA certification issued prior to the effective date of the final rule. While EPA's top priorities are to ensure that emission reductions occur in a timely manner and that there is no backsliding from the improvements that many manufacturers have already made, we have also sought to avoid unreasonable economic impacts on those manufacturers (over 95% of which are small businesses) who need additional time to develop, test, field evaluate, and certify a full range of cleaner models across their consumer product lines. In 1988, there were "logjam" concerns about the capacity of accredited laboratories to conduct certifications tests and the time for the EPA to review the tests and adequately assure compliance if all the NSPS requirements were to be immediate. Those concerns have been expressed this time also. The proposed phased implementation approach would help reduce those concerns. We ask for specific comments on the length of this proposed transition and the degree to which there would be any critical economic impacts on manufacturers who have heaters with current certifications if we were to allow less than the full 5-year certification period for model lines certified prior to the effective date of this rule but the heaters are manufactured after the effective date of the final rule.

We are proposing a two-step, phased implementation approach (referred to herein as the "Proposed Approach") that would apply to all new adjustable burn rate wood heaters, single burn rate wood heaters and pellet heaters/stoves required to comply with the Step 1 emission limits specified in the final rule. Under today's Proposed Approach, the Step 1 emision limits would apply to each heater (1) manufactured on or after the effective date of the final rule or (2) sold at retail on or after 6 months after the effective date of the final rule. Step 2 emission limits would apply to each adjustable rate wood heater, single burn rate wood heater and pellet heater/stove manufactured or sold 5 years after the effective date of the final rule. We ask for specific comments on the Proposed Approach and the degree to which these dates could be sooner.

We are also asking for comments on an alternative three-step approach (referred to herein as the "Alternative Approach") for all adjustable rate wood heaters, single burn rate wood heaters and pellet heaters/stoves.. Under this Alternative Approach, Alternative Step 1 emission limits would apply to each heater (1) manufactured on or after the effective date of the final rule or (2) sold at retail on or after 6 months after the effective date of the final rule. The emission levels for Step 1 and Alternative Step 1 are identical. Alternative Step 2 emission limits would apply to each heater manufactured or sold on or after the date 3 years after the effective date of the final rule.. Alternative Step 3 emission limits would apply to each heater manufactured or sold on or after the date 8 years after the effective date of the final rule. The Proposed Approach Step 2 emission limits and the Alternative Approach Step 3 emission limits are identical. We ask for specific comments on this Alternative Approach and the degree to which these dates could be sooner.

While the 1988 promulgated subpart AAA (53 FR 5860, February 26, 1988) included an additional 1-year compliance extension for low volume manufacturers, i.e., companies that manufacture (or export to the U.S.) fewer than 2,000 heaters per year, this proposal does not include such a compliance extension. We are not proposing to extend this delay to adjustable burn rate wood heaters or pellet heaters/stoves, because most of these appliances already meet the proposed Step 1 emission levels. See section V.C. of the preamble for more discussion of this topic. However, we are requesting comments on the possible need for such a compliance extension for single burn rate wood heaters, which are not subject to the current subpart AAA requirements.

Table 2-1 summarizes the compliance deadlines and PM emissions standards that would apply to each wood heater appliance under the Proposed Approach. Table 2-2 summarizes the compliance deadlines and PM emissions standards that would apply to each wood heater appliance under the Alternative Approach.

Table 2-1. Proposed Approach Subpart AAA Compliance Deadlines and PM Emissions Limits

Appliance	Compliance Deadlines	PM Emissions Limit
Adjustable Rate Wood Heaters or Pellet Stoves with Current EPA Certification Issued Prior to Publication of Final Rule	Transition period from 1988 rule through the later of publication of final revised rule or expiration of current certification (maximum of 5 years certification and no renewal)	4.1 g/hr for catalytic stoves and 7.5 g/hr for noncatalytic stoves
All Other Adjustable Rate Wood Heaters, Single Burn Rate Wood Heaters or Pellet Stoves (includes currently certified heaters after the certification expires, catalytic and noncatalytic)	Step 1: upon effective date of final rule) Step 2: 5 years after effective date n of final rule)	4.5 g/hr 1.3 g/hr

Table 2-2. Alternative Approach Subpart AAA Compliance Deadlines and PM Emissions Limits

Appliance	Compliance Deadlines	PM Emissions Limit
Adjustable Rate Wood Heaters or Pellet Stoves with Current EPA Certification Issued Prior to Publication of Final Rule	Transition period from 1988 rule through the later of publication of final revised rule or expiration of current certification (maximum of 5 years certification and no renewal)	4.1 g/hr for catalytic stoves and 7.5 g/hr for noncatalytic stoves
All Other Adjustable Rate Wood Heaters, Single Burn Rate Wood Heaters or Pellet Stoves (includes currently certified heaters after the certification expires, catalytic and noncatalytic)	Step 1: upon effective date of final rule Step 2: 3 years after effective date of final rule) Step 3: 8 years after effective date of final rule)	4.5 g/hr 2.5 g/hr 1.3 g/hr

We are proposing to have a single determination of BSER for both catalytic and noncatalytic heater systems. As in 1988, the EPA again considered requiring catalyst replacement on a regular schedule but determined that enforcement of such a requirement would be difficult. As before, we are proposing to require manufacturers to provide warranties on the catalysts and prohibit the operation of catalytic stoves without a catalyst. In addition, we are proposing to require warranties for noncatalytic stoves. We are not proposing efficiency standards at this time, however, we are proposing to require testing and reporting of these data.

We are also proposing to require emission testing and reporting based on both crib wood and cordwood for Step 1, while allowing manufacturers to choose whether to certify with crib wood or cordwood for Step 1. For Step 2, we are requiring certifying with cordwood only. "Crib wood" is a specified configuration and quality of dimensional lumber and spacers that was intended to improve the repeatability of the test method required in the current Residential Wood Heaters NSPS promulgated in 1988. "Cord wood" is a different specified configuration and quality of wood that is intended today to more closely resemble what a typical homeowner would use.

Our current data for CO emissions performance and methods of control are not sufficiently robust to support strong CO emission limits, and it would delay the NSPS if we were to seek additional data elsewhere at this time to support strong CO emission limits. Although we lack sufficient data to propose a separate CO emissions standard at this time, we propose to require that the manufacturer determine CO emissions during the compliance test and report those results to the EPA. We specifically request emission and cost data for systems that reduce CO emissions. If those systems warrant inclusion in the determination of BSER, we would consider doing so. Also, we ask for specific comments on whether we should require indoor CO monitors as a critical safety component for heaters installed in occupied buildings or other buildings or enclosures in which the operator would enter to add fuel to the heater or conduct other normal operation and maintenance of the heater.

Like the current subpart, the EPA is using its authority under Section 114 of the CAA to require each manufacturer to submit certifications of compliance with this rule for all models and all units. As in the 1988 rule, provided that the certifications are timely, complete, and accurate, the EPA will allow certification of compliance with the PM emissions standards to be determined based on testing of a representative unit within the model line rather than testing every unit. As in 1988, the cost of testing each unit would be an order of magnitude greater than the cost of a wood stove and would be economically prohibitive. Also, as in 1988, the testing of each unit could create a potential "logjam" that would stymie the certification of cleaner model lines. We recognize there is some concern that testing laboratories may not be able to meet the demand for certification tests in the first few years. However, the availability of additional ISO-accredited labs, the advance notice that industry has had concerning the NSPS prior to this proposal, and the time between this proposal and the proposed implementation deadlines of the final rule, should ensure that adequate compliance certification resources are available. In addition, to further respond to the concern regarding availability of testing laboratories, the proposed subpart expands the definition of "Accredited Test Laboratory" from just EPA-accredited laboratories to allow laboratories accredited by a nationally recognized accrediting body to perform testing for each of the test methods specified in this NSPS under ISO-IEC[3] Standard 17025 to conduct the certification testing. The laboratories would have to register their credentials with the EPA and report any changes in their accreditation and any deficiencies found under ISO 17025. The EPA would review that information to approve (or deny or revoke) the accreditation for the purpose of the determining compliance with the NSPS prior to the lab conducting any certification testing or related work used as a basis for compliance with this rule.

To ensure a practical, orderly transition, the proposal retains the current "Administrator Approval Process" to review the certification application, including test results, for the first year following publication of the final rule. At that point or earlier if chosen by manufacturers, the revised subpart would implement a "Certifying-Body-Based Certification Process." Under this process, after testing is complete, a certification of conformity with the PM emissions standards must be issued by a certifying body with whom the manufacturer has entered into contract for certification services. Similar to the lab requirements, the certification body would have to be accredited under ISO-IEC Standard 17065 and register their credentials with EPA and report any changes in their accreditation and any deficiencies found under ISO 17065. The EPA would review that information to approve (or deny or revoke) the accreditation for the purpose of the determining compliance with the NSPS prior to the lab conducting any certification testing or related work used as a basis for compliance with this rule. Upon review of the test report and quality control plan submitted by the manufacturer, the approved certifying body may certify compliance and submit the required documentation to the EPA's Office of Enforcement and Compliance Assurance for review, approval and listing of the certified appliance.

As in the 1988 NSPS, each affected unit would be required to have an applicable permanent label and have an owner's manual that contains specified information. We are proposing that permanent labels be required for each affected unit effective on the date of publication of the final rule. We proposing to no longer require showroom temporary labels ("hangtags") for each affected unit. This is a change from the existing 1988 NSPS, which requires that all certified models be equipped with temporary hangtags. The intent of the 1988 temporary hangtag requirement was to highlight models that met the EPA standards. We believe

---

[3] ISO, the International Organization for Standardization, and IEC, the International Electrotechnical Commission, prepare and publish international standards.

adequate information on EPA certifications would be available on the EPA Burn Wise website, the permanent label and the owner's manual. The proposal would clarify that the permanent label must be installed so that it is readily visible both prior to and after the unit is installed. This clarification is needed to document the use of complying heaters required by state and local rules and/or to determine the unit's applicability to any future changeout programs.

We request specific comments on how to best assure that manufacturers and retailers and online marketers of wood heaters only use valid certification test data and that regulators and consumers have ready access to certification information. We request specific comments on ways to improve the delivery of information and on whether different information might be useful to the consumer and to the regulatory authorities. We also request specific comments on what information and format might be most useful for the EPA to include on the EPA Burn wise website, EPA's web portal for information on residential wood smoke emissions and ways to reduce them, and wood burning appliances.

In addition to the PM emissions standards, certification and labeling requirements, we are proposing to continue to require the proper burn practices that already apply to the owner or operator of a wood heating appliance. That is, the 1988 standards already include the requirement that the owner or operator must operate the heater consistent with the owner's manual and not burn improper fuels. The proposed revision clarifies that the current requirement to operate according to the owner's manual must continue to include a list of prohibited fuel types that create poor or even hazardous combustion conditions and must include the direction that pellet fuel appliances can be safely and effectively operated only with pellet fuels used in the certification tests. We propose that pellets for the certification tests be only those that have been produced under a licensing agreement with the Pellet Fuels Institute (PFI), or equivalent (after request and subsequent approval by the EPA), to meet certain minimum requirements and procedures for a quality assurance process.[4] (Currently, PFI is the only organization that has volunteered to conduct such a program.) We believe that these provisions are necessary to ensure that the appliances operate properly such that emissions are reduced as intended. We ask for specific comments on whether we should include other requirements of best burn practices or adjustments to help ensure proper operation, *e.g.*, chimney height and draft specifications, moisture content of wood, and limits on visible emissions.

---

[4] Details of the PFI program are available at http://pelletheat.org/pfi-standards/pfi-standards-program/.

The proposed subpart still contains the crucial quality assurance provisions in the 1988 NSPS. The 1988 NSPS requirements for manufacturer quality assurance programs would be maintained for 1 year following the effective date of the final rule. At that point, the manufacturer would be required to adopt a Certifying-Body-Based Quality Assurance program. The Certifying Body would conduct regular, unannounced audits to ensure that the manufacturer's Quality Control Plan is being implemented properly.

The concepts of the EPA selective enforcement and random audit testing programs of the current 1988 NSPS will be retained under the Proposed rule, although they will be streamlined and simplified to better ensure compliance and to clarify that enforcement audits can be based on any information the EPA has available and do not have to be statistically random. Also, we clarify that the EPA and states are allowed to be present during the audits and that states may provide the EPA with information to help the EPA compliance assurance efforts.

The EPA is proposing a number of revisions to certification testing for various appliances. The EPA is proposing that updated and tailored versions of Method 28, a sampling and analysis method to analyze wood stove emissions, be used for all of the appliances in this rulemaking. The EPA developed Method 28 in 1987 and 1988 as part of our efforts on the 1988 NSPS. The manufacturers, laboratories, states, and the EPA have now had over 25 years of experience with Method 28 and it has been very useful for certifying hundreds of model lines of wood stoves. We asked the manufacturers, EPA-accredited laboratories, and states for their insights on Method 28. Many stakeholders agree that changes should be made to improve the reproducibility and repeatability of the test procedures and to address concerns about how to best ensure protection across the entire U.S. when various operating scenarios are used and various wood species and densities are used. For example, to address some of these concerns, ASTM, formerly known as the American Society for Testing and Materials, has used a "consensus-based" process to develop E2515-10 "Standard Method for Determination of Particulate Matter Emissions in a Dilution Tunnel." As with all test methods, there are opportunities for continual improvement, and the EPA requests specific comments and supporting data for additional potential improvements to E2515-10.

A number of states have expressed concern about ASTM's Intellectual Property Policy which requires all participants to give their intellectual property rights to ASTM so that, in turn, ASTM can control distribution of the drafts and final test methods and sell the final test methods to potential users. Attorneys General for several states have indicated that state employees in their states cannot give to ASTM the property rights for property that their states paid for via the employee salaries and other expenditures and thus cannot participate in ASTM's "consensus-based" process. For this rulemaking, ASTM is allowing public review, for no charge, of the ASTM test methods and draft work products relevant to this proposed rule at www.astm.org/epa. The EPA requests specific comments and supporting data on the substance of all of the test methods relevant to this rulemaking and specific comments on the ASTM process and ways to ameliorate the process concerns.

ASTM methods E2779-10 "Standard Test Method for Determining Particulate Emissions from Pellet Heaters" and E2780-10 "Standard Test Method for Determining Particulate Emissions from Wood Heaters", which are test methods used to determine average emissions rates and average emissions factors for pellet heaters/stoves and wood heaters/stoves, respectively, could potentially replace the wood heater fueling and operation requirements in Method 28 for these heaters. Note that ASTM intends to use the same E2515-10 for the sampling and analysis portion for all the appliances and then separate methods per appliance types for the fueling and operation portions of these methods. The EPA believes that E2779-10 is a sound method for measuring emissions from pellet stoves and includes reasonable measures to reduce testing costs for continuously-fed appliances and today we are proposing its use. However, because, as noted earlier, some states were not able to participate in the ASTM method development process, we specifically request comments and supporting data of all aspects of not only this test method but also all the proposed methods as part of the comments on today's proposed rule.

Similarly, the EPA believes that ASTM Method E2780-10 includes improvements for testing adjustable and single burn rate wood heaters, and we are proposing many of those improvements. For example, we are proposing the use of the E2780-10 appendix for testing single burn rate appliances. However, we, and some states, do not agree with all the changes that ASTM has made for adjustable burn rate wood heaters, and some provisions are not as protective as we, and some states, now believe they need to be. As noted above, several states are concerned about how to best ensure that the methods are protective for the entire U.S., considering differences in wood species, density, and homeowner operation. The EPA and the states are particularly concerned about scenarios in which stoves will have higher emissions in homes than the emissions measured in the laboratories. For example, the states and the EPA are concerned about the ASTM changes on burn rate categories, *i.e.*, easing or eliminating the lowest burn rates that often occur in home operations and are typically the dirtiest and least efficient. The EPA is asking for specific comments on these issues and recommendations and supporting data for other changes.

In addition, ASTM has developed a draft test method that uses cordwood rather than crib wood to better represent real world conditions. All stakeholders agree that a test method that better represents real world conditions would be a significant improvement and help ameliorate concerns that some heaters do not perform as well in the field as they do in laboratories. We are also interested in real-time emission test methods that measure cold or warm start-up emissions and emission peaks/durations. We are also interested in field test methods and less expensive test methods that regulators and neighbor can use to better quantify impacts in the real world. The EPA is asking for specific comments and data on all these potential methods, issues and recommendations.

2.3   Central Heaters: Hydronic Heaters and Forced-Air Furnaces

The proposed subpart QQQQ would apply to new wood-fired residential hydronic heaters and forced-air furnaces and any other affected appliance as defined in proposed subpart QQQQ as a "central heater." These appliances are described in more detail in Section 3 of this RIA. We believe this "central heater" categorization will ensure that all appliances potentially affected under the proposed subpart QQQQ are properly included. The proposed provisions of subpart QQQQ would apply to each affected unit that is manufactured or sold on the effective date of the final rule. This proposal does not include any requirements for heaters that are fueled solely by gas or oil or coal or non-wood biomass. In addition, this proposal does not include any requirements associated with wood heaters that are already in use. The EPA continues to encourage state, local, tribal, and consumer efforts to change out (replace) older heaters with newer, cleaner, more efficient heaters, but that is not part of this Federal rulemaking.

The Proposed Approach (or Option) would apply to new residential hydronic heaters and forced-air furnaces. Under the Proposed Approach, the Proposed Step 1 emission limit would be upon the effective date of the final rule. The Proposed Step 2 emission limit would be 5 years after the effective date of the final rule. We ask for specific comments on the Proposed Approach and the degree to which these dates could be sooner.

We also considered an alternative three-step approach (Alternative Approach or Option) for residential hydronic heaters and forced air heaters. As in the Proposed Approach, under this Alternative Approach, the Alternative Step 1 emission limit would be upon the effective date of the final rule. The Proposed Step 1 emission limit and the Alternative Approach Step 1 emission limit are identical. The Alternative Step 2 emission limit would be 3 years after the effective date of the final rule. This serves as an "interim" step on the way to the tighter emissions limits included in Alternative Step 3. The Alternative Step 3 emission limit would be 8 years after the effective date of the final rule. The Proposed Step 2 emission limit and the Alternative Approach Step 3 emission limit are identical. We ask for specific comments on this Alternative Approach and the degree to which these dates could be sooner.

Table 2-3 summarizes the proposed compliance dates and PM emissions standards that would apply under the Proposed Approach. Table 2-4 summarizes the compliance dates and PM emissions standards that would apply under the Alternative Approach. Similar to subpart AAA, we are not proposing a standard for CO or efficiency, but are proposing to require manufacturers to collect and report CO emissions and efficiency data during certification tests.

Table 2-3. Proposed Approach Subpart QQQQ Compliance Dates and PM Emissions Standards

Appliance	Compliance Date	Particulate Matter Emissions Limits
Residential Hydronic Heater	Step 1:Upon effective date of the final rule)	0.32 lb/MMBtu heat output and a cap of 7.5 g/hr for individual test runs
	Step 2: 5 years after effective date of the final rule)	0.06 lb/MMBtu
Forced-Air Furnace	Step 1: Upon effective date of the final rule)	0.93 lb/MMBtu
	Step 2: 5 years after effective date of final rule)	0.06 lb/MMBtu

Table 2-4. Alternative Approach Subpart QQQQ Compliance Dates and PM Emissions Standards

Appliance	Compliance Date	Particulate Matter Emissions Limits
Residential Hydronic Heater	Step 1:Upon effective date of the final rule)	0.32 lb/MMBtu heat output and a cap of 7.5 g/hr for individual test runs
	Step 2: (3 years after effective date of final rule)	0.15 lb/MMBt
	Step 3: 8 years after effective date of the final rule)	0.06 lb/MMBtu
Forced-Air Furnace	Step 1: Upon effective date of the final rule)	0.93 lb/MMBtu
	Step 2: 3 years after effective date of final rule)	0.15 lb/MMBtu
	Step 3: 2022 (8 years after	

publication of final rule)	0.06 lb/MMBtu

Unlike the subpart AAA requirements, the subpart QQQQ requirements would not provide an additional time period for the sale of unsold units manufactured before the compliance date nor do they include a small volume manufacturer compliance extension.[5] We ask for comments on the timing for implementation.

As in the current subpart AAA for wood heaters/stoves, we are proposing a list of prohibited fuels because their use would cause poor combustion or even hazardous conditions. We request comment on these requirements and data to support additional requirements, if warranted. Also, as in the current subpart AAA for wood heaters/stoves, we are proposing that the owner or operator must not operate the hydronic heater or forced-air furnace in a manner that is inconsistent with the owner's manual. For pellet-fueled appliances, the proposal makes it clear that operation according to the owner's manual includes operation only with pellet fuels that have been used in the certification test and have been graded and marked under a licensing agreement with the PFI, or equivalent (after request and subsequent approval by the EPA), to meet certain minimum requirements and procedures for a quality assurance process. Details of the PFI program are available at http://pelletheat.org/pfi-standards/pfi-standards-program/. (Currently, PFI is the only organization that has volunteered to conduct such a program.) We believe that these provisions are necessary to ensure that the appliances operate properly such that emissions are reduced as intended. We ask for specific comments on the use of the PFI program and the PFI specifications, especially the degree to which the PFI program will adequately ensure the absence of construction and demolition waste (and associated toxic contaminants) in pellets.

The proposed permanent labels and owner's manual requirements are similar to the guidelines in the EPA's current voluntary hydronic heater program with some improvements. We provide information on the number of models that currently meet the limits in the voluntary hydronic heater program in Section 4. We request specific comments on ways to improve the delivery of information on the permanent labels and in the owner's manual and the Burn Wise website and whether additional information might be useful to the consumers and to the regulatory authorities.

The structure of the rest of the proposed subpart QQQQ is similar to the proposed subpart AAA certification and quality assurance process. We request specific comments on changes or

---

[5] See section V.C. of the preamble for more discussion of this topic.

improvements to that process that might be needed to address any special concerns related to the certification of hydronic heaters and forced-air furnaces.

The EPA developed Method 28 OWHH (outdoor wood hydronic heaters) in 2006 as part of our efforts for voluntary qualification of cleaner hydronic heaters. We received input at that time from manufacturers, laboratories, and some states in order to quickly develop a mostly consensus-based method that we incorporated into the program partnership agreements. We used Method 28 for wood stoves as the foundation, and thus, Method 28 OWHH has many aspects similar to Method 28. Three significant differences are (1) Method 28 OWHH uses larger cribs because hydronic heater fireboxes are typically much larger than wood heater fireboxes, (2) Method 28 OWHH uses red oak instead of Douglas fir because red oak is the more common fuel in the U.S., and (3) Method 28 OWHH includes procedures for determining 8-hour heat output and efficiency. The manufacturers, laboratories, states, and the EPA have now had over 7 years of experience with Method 28 OWHH and its successor Method 28 WHH (wood hydronic heaters, improved and expanded to include indoor heaters, not just outdoor heaters).

All the stakeholders that have provided input on the test methods agree that the methods should be thoroughly vetted and changed as necessary to improve the methods' accuracy and precision and to address concerns about how to best ensure real world protection across the entire U.S. when various operating scenarios and wood species and densities are used. ASTM has developed E2618-09, a test method that applies to wood-fired hydronic heaters, to address some of these concerns, and the EPA believes that E2618-09 does include some improvements. However, as with the wood stove methods, we and some states do not agree with all the changes that ASTM has made. For example, the states of Washington and Oregon are very concerned that Method 28 WHH and ASTM E2618-09 do not specify fueling with Douglas fir, which is used in EPA Method 28 for wood stoves and which these states require in their regulations for residential wood heaters, including hydronic heaters, and is used frequently in their states for fuel in the real world. They are concerned that hydronic heaters tested with red oak will have higher emissions when fueled with Douglas fir and other less dense species typical in their states. Also, a number of states and the EPA are concerned about the ASTM changes to the burn rate categories, *i.e.*, easing or eliminating testing at the lowest burn rates which often occur in home operations and are typically the dirtiest and least efficient. For several years, we have been communicating with European certification laboratories to learn how they conduct their tests under EN 303-5, a European Union test protocol for wood-burning appliances, and to consider if incorporating some of their testing procedures might improve our test methods.

More recently, because of initial concerns about some surprisingly high laboratory test efficiencies for a couple of the EPA voluntary program Phase 2 qualified partial heat storage models, the EPA, the Northeast states that regulate hydronic heaters, laboratories (including EPA-accredited laboratories and Brookhaven National Laboratory), and manufacturers have conducted an exhaustive review of voluntary program qualifying test reports. All of the stakeholders that provided input on the test methods agree that we need a change in the test method for testing of non-integral partial heat storage models (*i.e.*, models that have separate heat storage but the storage does not have the capacity to safely handle all the heat generated by a full load of fuel). ASTM has been leading an effort to develop an Appendix X2 , which is additional guidance as support, to the test method for such models but has not completed that effort as of today's proposal. Brookhaven National Laboratory has recommended a method to the New York State Department of Environmental Conservation (NYSDEC) and NYSDEC is using that method for certification of such models in their state. EPA is proposing that method be used for certification of the NSPS for hydronic heaters equipped with a heat storage unit.[6]

Further, we are proposing revisions to Method 28 WHH that would require that all affected non-pellet hydronic heaters, subject to new subpart QQQQ, conduct and report certification testing using both crib wood and cordwood for the Step 1 emission limits and then choose which they want to use for compliance. For other than pellet-fueled heaters, the compliance tests would be solely cordwood for the Step 2 emission limits.

We are asking for specific comments on whether the EPA should use (1) one or more of the draft versions of Appendix X2 being considered as part of ASTM work product WK26581, which is a revision to the existing E2618-09 test method for measuring emissions from outdoor hydronic heaters; (2) the European Union test method EN 303-05 as the Maine Department of Environmental Protection approved for certification of hydronic heaters in their state as equivalent to the EPA Method 28 WHH; (3) the use of the NYSDEC partial thermal storage test method; and/or (4) some other test method. For use of any of the test methods, the EPA would require that the amount of heat storage for the actual sale and installation of the hydronic heaters be no less than the amount used for the certification tests. Because EN303-05 does not currently utilize heat storage during the certification test, if the EPA were to use EN303-05 test results, the EPA would require the installed heater to have heat storage that can safely handle at least 60% of the maximum heat output of the heater or a greater level if the manufacturer specifies a greater level. The EPA is asking for specific comments on the appropriateness of this heat storage level

---

[6] *See* footnote 19.

or other levels. The EPA will consider any or all of these options as the preferred reference test methods or as acceptable emission testing alternatives. (ASTM previously developed an Appendix X1, an additional part to the test protocol, for testing of models that have "full" heat storage that can safely accept the heat from the full load of fuel.) We request comments on all aspects of heater testing and are especially interested in emission test data that compare the results for testing by these different methods.

The exhaustive review discussed above found a number of areas in the methods to improve the quality of the data and reduce anomalies. In June 2011, the voluntary program stakeholders agreed to a number of changes to Method 28 OWHH, and we are proposing the revised method today as EPA Reference Method 28 WHH. The EPA is asking for specific comments on this method and recommendations and supporting data for other changes or acceptable alternatives. Additional information on the EPA methods is available at www.epa.gov/burnwise and the ASTM methods and draft work products are available at www.astm.org/epa.

As for wood heaters/stoves under Subpart AAA, ASTM is developing hydronic heater test methods that use cordwood instead of crib wood in order to better represent real world conditions. The proposed Step 2 of subpart QQQQ will require testing using cord wood. The EPA requests specific comments and data to support the ASTM cord wood methods and/or other cord wood test methods.

The EPA is proposing to rely on the test method B415.1-10 that has been developed by the Canadian Standards Association (CSA) for forced-air furnaces. All CSA standards are developed through a consensus development process approved by the Standards Council of Canada This process brings together volunteers representing varied viewpoints and interests to achieve consensus and develop a standard. CSA worked for years on development of this test method that has its roots in earlier U.S. efforts on wood stoves. The current version of CSA B415.1-10 was published in March 2010, and it includes not only the forced-air furnace test method but also new Canadian emission performance specifications for indoor and outdoor central heating appliances.

Although the CSA B415.1-10 technical committee included 32 individuals, including numerous U.S. manufacturers and laboratories, it did not include any states or environmental groups, and the EPA participation was minimal. Nevertheless, we are satisfied that this CSA method warrants proposal for this rulemaking and we request specific comments and supporting data. We ask for specific comments on the appropriateness of using the CSA test method in its entirety, including the use of cordwood instead of cribs that are used in current versions of Method 28 and Method 28 WHH. To review the CSA test method, please go to www.csa.ca.

2.4   New Residential Masonry Heaters

The proposed subpart RRRR would apply to new residential masonry heaters. A masonry heater is a site-built or site-assembled, solid-fueled heating device constructed mainly of masonry materials in which the heat from intermittent fires burned rapidly in its firebox is stored in its massive structure for slow release to the building. It has an interior construction consisting of a firebox and heat exchange channels built from refractory components.. We are proposing that, as of the effective date of the final rule, no person would manufacture or sell a residential masonry heater that does not meet the proposed emission limit of 0.32 lb of PM per MMBtu heat output. We are also proposing a 5-year small volume manufacturer compliance extension that would apply to companies that construct fewer than 15 masonry heaters per year. See section V.C. of the preamble for more discussion of compliance date related issues. We request specific comments on the degree to which these dates can be sooner. As in the case of the other proposed standards, we are proposing requirements that would apply to the operator of the masonry heater, including a provision to operate the unit in compliance with the owner's manual; a prohibition on use of certain fuels; and a requirement to use licensed wood pellets or equivalent, if applicable. We are not proposing efficiency standards for new residential masonry heaters at this time because data are not yet available to support the basis for such standards. As in the case of the other proposed standards, this masonry heaters proposal does not include any requirements for heaters that are fueled solely by gas or oil or coal or non-wood biomass. Also as in the case of the other proposed standards, this masonry heaters proposal does not include any requirements associated with heaters already in use.

The EPA is proposing to rely on ASTM test method E2817-11. The laboratories, some states, and many in the masonry heater industry worked for years on drafts of this method that has its roots in earlier regulatory efforts in Colorado. The EPA has participated in the discussions from time to time over the years and has provided comments and suggestions. There have been a number of variations of similar methods over the years. The current ASTM drafts are ASTM E2817-11 "Standard Test Method for Test Fueling Masonry Heaters" and ASTM WK26558 "Specification for Calculation Method for Custom Designed, Site-built Masonry Heaters." (*see* http://www.astm.org/DATABASE.CART/WORKITEMS/WK26558.htm for method). We are encouraged by the progress shown by development of these current draft ASTM methods and propose that they be used for this rulemaking. We request specific comments on these draft methods and any changes that should be considered and supporting data for those changes. We request specific comments and supporting emission test data on the use of "Annex A1. Cordwood Fuel" and "Annex A2. Cribwood Fueling." ASTM is allowing public review, for no charge, of the ASTM test methods and draft work products relevant to this rule at www.astm.org/epa.

As an alternative to testing, we are proposing that manufacturers of masonry heaters submit a computer model simulation program, such as ASTM WK 26558 noted above for the EPA's review and approval.

The structure of the rest of the proposed subpart RRRR is similar to the proposed subpart AAA certification and quality assurance process and contains similar requirements for labels, owner's manual, etc. One difference, however, is that, for small custom unit manufacturers, we are requiring less stringent QA procedures. Specifically, we are proposing that the initial certification for these custom units is sufficient and that no further QA regulatory requirements are necessary because each unit is a unique model and subject to certification. We request comment on changes or improvements that might be needed to address special concerns related to certification of masonry heaters.

SECTION 3
INDUSTRY PROFILE

The proposed revisions to the NSPS for residential wood heaters would cover a number of devices that include wood stoves/heaters, pellet stoves/heaters; masonry heaters; indoor and outdoor hydronic heaters and forced-air furnaces. (This RIA and the proposal use the terms stove, heater, and stove/heater interchangeably.) EPA has developed this industry profile to provide the reader with a general understanding of the technical and economic aspects of the industries that would be directly affected by potential revisions to the NSPS regulation for new residential wood heaters and to offer information relevant to preparing an economic impact analysis (EIA) for this proposed revision to the NSPS. We begin by outlining the supply side by discussing the production process for wood heaters and the associated costs and follow this with an overview of the demand side of the market for residential wood heaters as a primary or secondary home heating system. We then address the characteristics that define the residential wood heating market and profile the companies that produce wood heating systems. Although the wood heating equipment industry includes multiple product markets, there is little published information about the intricacies of each individual market. For this profile, we analyzed the wood heating market primarily on an aggregated level and provide detailed information for specific product markets when such information is available.

3.1   Supply Side

Wood heating devices embody a variety of products that provide heat for residential consumers by burning wood or other solid biomass fuel. Indoor wood-burning devices can provide space heating for a single room or can be central heaters for a residential home. Indoor heating devices include freestanding wood stoves, pellet stoves, masonry heaters, fireplace inserts, and forced-air furnaces. Outdoor wood heating devices, also known as outdoor wood boilers, or water stoves, are typically located adjacent to the home they heat in small sheds with short smoke stacks. Other products considered in the development of potential proposed revisions of this NSPS (but not proposed to be regulated in this rulemaking) include low-mass fireplaces, open masonry fireplaces, fireplaces, fire pits, chimineas, cook stoves, and pizza ovens.

This section provides a general description of the residential wood heater manufacturing processes. We then provide more detailed definitions of the indoor and outdoor wood heater products considered and the wood fuels used in their operation.

3.1.1 Production Process

The manufacturing process for residential wood heaters varies depending on the product type being produced. Generally, the manufacturing process entails the assembly of several prefabricated metal components. Major inputs include cast iron, metal products, heat-proof glass, fireproof fabric insulation, refractory brick, and heat-tolerant enamels or coatings.

Wood heating devices are typically categorized by emissions and efficiency ratings. The emissions ratings typically use EPA, ASTM, CSA, or EN (European Union) test methods. The efficiency ratings are based on tests that measure the amount of heating value transferred from a full load of wood or other biomass fuel (fuel type varies based on the product being tested) to the living space. Efficiency tests evaluate two performance metrics that include combustion and heat transfer efficiency. Combustion efficiency determines how effective the fire box design is at burning the fuel and extracting its heating value. Heat transfer efficiency tests are potentially conducted in calorimeter rooms equipped with temperature sensors to measure the degree changes in the heated living space and the flue exhaust to determine how much heat from the fire is delivered to the living space compared with the heat lost up the flue (EPA, 2009c).

Thermal output, typically expressed in British thermal units per hour (BTU/hr) in the United States, is the heat output measure that tells the amount of heat produced each hour. A higher BTU/hr rate suggests that a stove will produce more heat per hour than a stove with a lower rating. Depending on design and size characteristics, a space heating device heat output rating ranges between 8,000 and 90,000 BTU/hr. Larger heating systems designed to provide whole home heating have heat output ratings that range from 100,000 to greater than one million BTU/hr.

3.1.2 Product Types

3.1.2.1 Wood and Coal Stoves

EPA-certified wood stoves typically are enclosed combustion devices that provide direct space heating for a specific room or area of a home.[7] Catalytic and noncatalytic wood stoves are two general types of wood stoves available in the United States. (Some models are hybrids.) This designation refers to the design of the combustion system. Noncatalytic combustion systems rely on high temperatures (>1,000°F) within the fire box to fully combust the chemical compounds (combustible gases and particles) in the wood smoke. In catalytic combustion systems, the

---

[7] EPA-certified wood stoves are those wood stoves that meet the requirements under the current residential wood heater NSPS.

presence of the catalytic element lowers the temperature at which wood smoke chemical compounds combust. Catalytic elements or combustion system designs in noncatalytic combustion systems are used in existing stoves to meet EPA emission standards.

Coal stoves are similar in structure and appearance to wood stoves. Most coal stoves are designed to burn hard anthracite coal instead of soft bituminous coal (Houck, 2009), but different varieties of coal have been used in coal stoves over time. Stoves that solely burn coal are not affected by the proposed revisions to the NSPS.

3.1.2.2 Wood Pellet Stoves and Biomass Stoves

Wood pellet and other biomass stoves are similar in application to wood stoves but generate heat through pellet combustion. Wood pellet stoves use tightly compacted pellets of wood as fuel, whereas other biomass stoves can use a variety of pellet types, including corn, fruit pits, and cotton seed (EPA, 2009c). A load of pellets is poured into the stove's hopper; then the user sets a thermostat that controls a feed device within the stove. The feed device regulates the amount of fuel that is released from the hopper into the heating chamber, which is where the combustion takes place (EPA, 2009c). Pellet stoves are typically more efficient in terms of combustion and heating than standard wood stoves but require electricity to operate the fans, controls, and pellet feeders (EPA, 2009c). Stoves that solely burn non-wood biomass are not affected by the proposed revisions to the NSPS.

3.1.2.3 Masonry Heaters

A masonry heater is a solid-fueled heating device that is pre-manufactured or constructed on site using mainly masonry or ceramic materials (Masonry Heater Association of North America, 1998). Though masonry heaters and traditional fireplaces are similar in appearance, masonry heaters are used primarily to generate heat, whereas fireplaces typically serve a more aesthetic purpose. The heater itself is made up of an interior construction unit consisting of a firebox and a set of heat exchange channels (Chernov, 2008). The hot gas produced during rapid combustion of fuel within the firebox passes through the heat exchange channels, which run throughout the structure and saturate the masonry mass with heat (Chernov, 2008). Most masonry heaters weigh over 800 kg. After the masonry walls are saturated, the masonry heater radiates the heat into the area for 12 to 15 hours (Chernov, 2008). Masonry heaters can heat a home all day without having to burn continuously and are often used in areas where other fuel sources are unavailable (Chernov, 2008). However, there is a significant lag time between the initial burn and the time that the masonry structure releases sufficient heat to warm a living space (U.S. Department of Energy [DOE], 2010).

3.1.2.4 Fireplace Inserts

A fireplace insert is a type of heater/stove that is designed to fit inside the firebox of an existing wood-burning fireplace (Wood Heat Organization, 2010). EPA-certified fireplace inserts are essentially wood heaters/stoves without legs or pedestals. An insert is made of steel or cast iron and is typically installed in masonry fireplaces or traditional fireplaces in order to provide effective heating (Hearth, Patio, and Barbeque Association [HPBA], 2010b). As an insulated closed-door system, a fireplace insert improves combustion by slowing down the fire, decreasing the excess air, and increasing the fire's temperature (HPBA, 2010b). In addition to wood-fueled fireplace inserts, other inserts can be fueled with natural gas, propane, pellets, or coal (HPBA, 2010b).

3.1.2.5 Forced Air Furnaces

A forced-air furnace is a type of central heating system that typically burns cordwood or pellets. A forced-air furnace is typically located inside a house and provides controlled heat throughout a home using a network of air ducts (EPA, 2009c). This is a primary heating system that requires electricity to operate and is much more common currently in the U.S. compared to hydronic heaters.

3.1.2.6 Outdoor Wood Heaters

An outdoor wood heater, also often called a wood-fired boiler, is a type of hydronic heater that is designed to be the home's primary heating system. Wood boilers are typically located outdoors and have the appearance of a small shed with a smokestack (EPA, 2009c). Hydronic heaters burn wood to heat a working liquid contained in a closed-loop system. The heated liquid is then circulated to the house to provide heat and hot water (EPA, 2009). Hydronic heaters are typically sold in areas with cold climates where wood may be the most readily available fuel source (EPA, 2009c). In addition to outdoor hydronic heaters, there is an emerging market for indoor hydronic heaters. Currently, the indoor hydronic heater market is approximately 10% of the hydronic heater market.

3.1.2.7 Indoor and Outdoor Fireplaces

Fireplaces are typically not effective heating sources and are typically considered more of an aesthetic feature than a functional device. The common low-mass fireplace is pre-fabricated of steel in a factory and shipped to the home builder. A low-mass fireplace and its attached chimney are light enough to be weighed on a platform scale (EPA, 2009c). Although low-mass fireplace installations in homes often surrounded by natural or synthetic facades of masonry–like materials, they should not be confused with masonry fireplaces or masonry heaters that are primarily constructed if brick, stone, or other masonry materials. Masonry fireplaces are traditional, aesthetic fireplaces that do not have the extensive heat channels that define masonry heaters (Fireplaces & Woodstoves, 2010).

Fireplaces are also often used to enhance the outdoor area of a house. A portable grated cylinder style has a bottom basin surrounded by open grating for a fire, a cooking grate, and a lid (EPA, 2009c). A permanent outdoor fireplace is similar to one that would be found indoors. They can be freestanding or attached to the outside of the house (EPA, 2009c).

Indoor and outdoor fireplaces are not covered by or affected by the proposed revisions to the NSPS.

3.1.2.8 Fire Pits, Chimineas, Cook Stoves, and Pizza Ovens

Several outdoor appliances involve using wood fuel for cooking or heating. A fire pit is a round outdoor hearth appliance that is designed to replicate the ambiance of a campfire by radiating heat in 360 degrees around the pit (HPBA, 2010c). A chiminea is typically constructed out of cast iron, terra cotta, or clay and burns firewood inside the internal oven. As the fire burns, the walls of the oven absorb heat. After the dome chamber reaches the desired temperature, the fire can be allowed to die down (EPA, 2009c). Wood cook stoves are made of cast iron to withstand the high temperatures produced by the fire (EPA, 2009c). They are similar in appearance to a conventional stove, complete with an oven and cooking ranges, but are larger in order to accommodate the wood fuel (EPA, 2009c). North American traditional cook stoves have defined dimensions and cooking performance characteristics. Native American bake ovens have defined cultural and cooking functions. Pizza ovens are made out of a masonry material, such as clay adobe or refractory bricks, which can endure high temperatures for an extended period of time (EPA, 2009c).

These appliances are not covered or affected by the revisions to the NSPS.

3.1.3   Costs of Production

Because of the variety of products covered under the wood heat source category, different manufacturers use a wide range of materials and have varying labor requirements. Since there is significant diversity in output between the producers in this category, as well as the broader industries in which they may be classified for data purposes, this section highlights the production costs associated with several of the North American Industry Classification System (NAICS) codes under which a significant number of the wood heating equipment manufacturing facilities in our database are included.

Table 3-1 displays costs for the heating equipment and hardware manufacturing industries. The production of devices like wood stoves, hydronic heaters, and fireplace inserts is included under the heating equipment category (NAICS 333414). In 2011, the total cost of materials used for production represented roughly 47% of the industry's total value of shipments,

Table 3-1. Costs for Labor and Materials for U.S. Heating Equipment and Hardware Manufacturing: 2011

NAICS-Based Code	Meaning of NAICS-Based Code	Year	Number of Employees	Annual Payroll ($1,000)	Production Workers Average per Year	Total Cost of Materials ($1,000)	Materials, Parts, Containers, Packaging, etc., Used ($1,000)	Cost of Purchased Fuels ($1,000)	Cost of Purchased Electricity ($1,000)	Total Value of Shipments ($1,000)
333414	Heating Equipment (except warm air furnace)	2011	15,925	803,254	9,497	1,968,956	1,639,894	10,840	41,594	4,153,470
332510	Hardware Manufacturing	2011	24,406	1,175,743	16,956	3,362,268	2,731,577	21,592	52,530	6,256,338

Source: U.S. Census Bureau. 2011a. American Fact Finder. Sector 31: Annual Survey of Manufactures: General Statistics: Statistics for Industry Groups and Industries: 2011 and 2010. http://factfinder.census.gov. Accessed on February 21, 2013.

while labor costs (represented as annual payroll estimates) only represented 19%. The hardware manufacturing industry (NAICS 332510) had similar statistics: materials used and annual payroll accounted for 54% and 19% of the total value of shipments, respectively.

Masonry fireplace construction and other site-assembled fireplace construction are covered under the two industries displayed in Table 3-2. The new single-family construction general contractors industry (NAICS 236115) covers a broad spectrum of construction activities beyond masonry and fireplace construction. Like heating equipment and hardware manufacturing, this industry is highly capital intensive; 42% of the value of the construction work is attributed to the cost of materials. Labor accounts for only 12%. For the masonry contractor industry, however, payroll costs represent over 30% of the value of construction work suggesting that masonry contracting requires a special skill set and a specific degree of craftsmanship.

The 2007 production costs for plumbing and heating equipment wholesalers (NAICS 423720), which are the most recent available from the Census Bureau, are outlined in Table 3-3. This category, which includes the merchant wholesale production of cooking and heating stoves and hydronic heaters, made over $50 billion in sales in 2007. Table 3-4 displays the costs for certain home furnishing stores, including those that sell wood stoves at retail prices. The costs for these industries may be more indicative of the wholesale and retail exchanges of wood-heating equipment rather than the actual production process.

3.2   Demand Side

The subject wood-fired heaters are sold explicitly for residential use. These devices can be included in the original construction of a new home or installed later in the life of the home. Demand for residential wood heating devices is driven by several key factors that include size, price, efficiency, aesthetics, and fuel type (e.g., cord wood, pellet wood, or other biomass fuels). However, consumer demand for any one product discussed in Section 3.1 is driven primarily by the intended end-use heating application. This section defines the three major consumer segments that drive demand based on the end-use application. Following this discussion, we present some national statistics on the variation in residential wood heat consumers in the United States. We conclude our discussion of the demand side by characterizing some of the substitutes for residential wood-burning devices.

Table 3-2. Costs for U.S. Masonry Contractors and Single-Family Home Contractors: 2007

NAICS-Based Code	Meaning of NAICS-Based Code	Year	Number of Employees	Total Payroll ($1,000)	Cost of Materials, Components, and Supplies ($1,000)	Total Value of Construction Work ($1,000)
236115	New single-family general contractors	2007	259,905	10,834,064	37,676,878	89,282,708
238140	Masonry contractors	2007	232,315	8,250,581	8,594,565	26,984,381

Source: U.S. Census Bureau. 2010b. American Fact Finder. Sector 23: EC0723SG01: Construction: Summary Series: General Summary: Detailed Statistics for Establishments: 2007. Released May 18, 2010. http://factfinder.census.gov.

Table 3-3. Costs for U.S. Plumbing and Heating Equipment Supplies Wholesalers: 2007

NAICS-Based Code	Meaning of NAICS-Based Code	Year	Number of Employees	Annual Payroll ($1,000)	Operating Expenses ($1,000)	Sales ($1,000)
423720	Plumbing and heating equipment supplies (hydronics) merchant wholesalers	2007	87,907	4,542,337	8,311,462	50,316,133

Source: U.S. Census Bureau, 2010c. American Fact Finder. Sector 42: EC0742A1: Wholesale Trade: Geographic Area Series: Summary Statistics for the United States, States, Metro Areas, Counties, and Places: 2007. Released July 23, 2010. http://factfinder.census.gov.

Table 3-4. Costs for U.S. Specialized Home Furnishing Stores: 2007

NAICS-Based Code	Meaning of NAICS-Based Code	Year	Number of Employees	Annual Payroll ($1,000)	Sales ($1,000)
442299	All other home furnishing stores	2007	19,057	3,427,682	27,326,976

Source: U.S. Census Bureau, 2010d. American Fact Finder. Sector 44: EC0744A1: Retail Trade: Geographic Area Series: Summary Statistics for the United States, States, Metro Areas, Counties, and Places: 2007. Released July 23, 2010. http://factfinder.census.gov.

3.2.1   End-Use Consumer Segments

The intended end-use heating application is a primary driver of demand for residential heating devices. The U.S. Annual Housing Survey (HUD, 2008) provides a starting point for classifying the various types of residential consumer of heating equipment. For the purposes of this profile we grouped consumers into three major segments based on their desired heating needs: whole-house heating, secondary or zone heating, and recreational outdoor heating applications.

The primary, or whole-house heating segment, includes homes with no other central heating system that can provide heating service in or outside the house. In smaller homes, a large stove or masonry heater may be sufficient to provide heat to the entire house. However, larger homes typically require, either individually or in some combination thereof, an outdoor wood boiler, a hydronic heater, or a pellet-burning forced-air furnace, to meet the consumers' heating needs.

The secondary, or zone heating segment, includes consumers that desire supplemental heat from a wood-burning device in homes with an existing central heating system that serves as the home's primary heat source. Cordwood and wood pellet-burning stoves are ideal for heating a single room or zone within a home. Smaller masonry heaters are also well suited for zone heating needs.

Finally, a third component of demand is represented by consumers who desire a wood-burning device for recreational aesthetic applications. Outdoor fireplaces, chimineas, outdoor ovens, and pizza stoves are some examples of the wood-burning devices designed for recreational applications. The products that address the needs of this consumer segment are primarily intended to enhance the aesthetics or landscape outside the home. Indoor fireplaces typically serve aesthetic or recreational purposes rather than providing effective room heat. Only about 9% of wood fireplaces are used for heat generation (HPBA, 2010a).

3.2.2   Regional Variation in Residential Demand

In 2010, 2.1% of total occupied homes in the United States relied on wood heat as the primary fuel source for home heating. The demand for wood heat is concentrated in the Northeast, the Northwest, and the northern Midwest regions of the United States. Table 3-5

Table 3-5. Wood as Primary Fuel Source for Home Heating in the United States: 2006–2008

State	Percentage of State Owner-Occupied Houses	Percentage of National Owner-Occupied Houses	Count
California	2%	9%	165,440
New York	3%	6%	103,740
Pennsylvania	3%	6%	100,355
Washington	6%	5%	92,664
Michigan	3%	5%	85,712
Wisconsin	5%	5%	83,040
Oregon	8%	4%	79,637
Ohio	2%	4%	67,665
Virginia	3%	3%	60,579
North Carolina	2%	3%	58,397
Minnesota	3%	2%	43,234
Maine	10%	2%	41,509
Indiana	2%	2%	38,550
West Virginia	7%	2%	38,142
Idaho	8%	2%	32,817
Colorado	2%	2%	28,668
Vermont	15%	1%	26,601
Massachusetts	2%	1%	25,870
Montana	9%	1%	24,355
New Hampshire	7%	1%	24,071
Top 20 total		68%	1,221,046
National total		100%	1,792,741

Source: U.S. Census Bureau. 2009. *American Community Survey: 2006–2008.* Available at:
http://factfinder.census.gov/servlet/DatasetMainPageServlet?_program=ACS&_submenuId=&_lang=en&_ts=.

illustrates this regional concentration by listing the 20 states that represent the highest percentage of households that use wood heat based on Census data from 2006–2008. The second column shows the number of wood-heat users as a percentage of the total homes in the state, while the third column shows the number of wood-heated homes as a percentage of the total users in the United States. These 20 states account for over two-thirds of the total primary U.S. residential wood heat demand.

About 10–12% of American households use wood when secondary wood heat demand is counted, according to the Census and the Energy Information Administration (EIA). Table 3-6 illustrates the regional breakdown of secondary wood heat demand by U.S. Census divisions in 2009, which is the most recent year for which these data are available. Roughly 8% of the American households use wood as a secondary heat source.

Table 3-6.   Wood as Secondary Heat Source by Census Division, 2009 (millions of households)

Census Division	Number of Households	Percentage of U.S. Households using Wood as Secondary Heat Source	Percentage of Total U.S. Households
South Atlantic	1.6	18%	1%
Pacific	1.7	19%	1%
East North Central	1.2	14%	1%
West North Central	0.8	9%	1%
West South Central	0.7	8%	1%
Mountain	0.8	9%	1%
Middle Atlantic	1.0	11%	1%
East South Central	0.4	4%	1%
New England	0.7	8%	0%
Grand total	8.8	100%	8%
Total U.S. households	113.6		

Source: U.S. Energy Information Administration, 2011. *Residential Energy Consumption Survey: 2009.* Available at http://www.eia.gov/consumption/residential/data/2009/#undefined.

Figure 3-1 shows which states fall into which Census divisions. More households rely on wood fuel as a supplemental heat source rather than as a primary source. Roughly 12% of American households used wood fuel for a secondary heat source in 2009, whereas 3% of households relied on wood for their primary heat source in the same year. The proportion of the population using primary wood heat was relatively consistent between 2005 data presented in Table 3-6 and the 2006 to 2008 period, as shown in Table 3-5.[4] One interesting note about

---

[4] Although the total occupied households between the Department of Energy's *Residential Energy Consumption Survey* [RECS] and the Census Bureau's *American Community Survey* [ACS] differ, the proportion of total occupied households using wood fuel as their primary home heating fuel is consistent. The survey data sources used in Table 3-6 assumes 111 million occupied homes in 2009 while Figure 2-2 assumes 109 million for the same year.

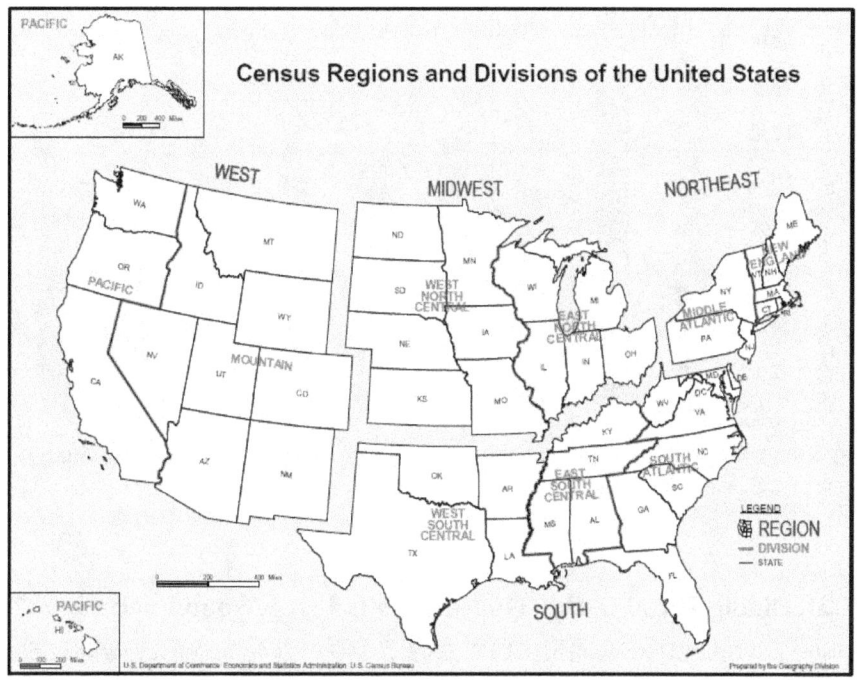

Figure 3-1.   Census Regions and Divisions of the United States

Source: U.S. Census Bureau, 2010e. *Census Regions and Divisions of the United States.* Available at
http://www.census.gov/geo/www/us_regdiv.pdf.

secondary wood fuel use is that it does not appear prevalent in the Middle Atlantic or New
England states, which account for only 19% of the total secondary use. This fact is in contrast to
the primary use data in Table 3-5, which shows households in Vermont, New Hampshire, Maine,
Pennsylvania, Massachusetts, and New York accounting for 17% of the total national primary
demand.

Within the wood heat demand constituency, there is also regional demand variation for
different wood-fueled appliances. For example, the demand for wood-fired forced-air furnaces is
concentrated primarily in the Great Lakes region of the country and, to a lesser extent, the
Midwest (HPBA, 2010a). These two regions account for 82% of the 30,000 to 35,000 furnaces
sold annually in the United States (HPBA, 2010a). Demand for wood-fueled cook stoves is
concentrated in the Amish and Mennonite communities in the Midwest (HPBA, 2010a).

3.2.3   National Home Heating Trends

Residential demand for wood fuel has been declining steadily throughout the United
States over a fairly long period of time. Figure 3-2 illustrates the number of households from
1989 to 2005 that reported using wood fuel for heating, cooking, or heating water. In 1989,

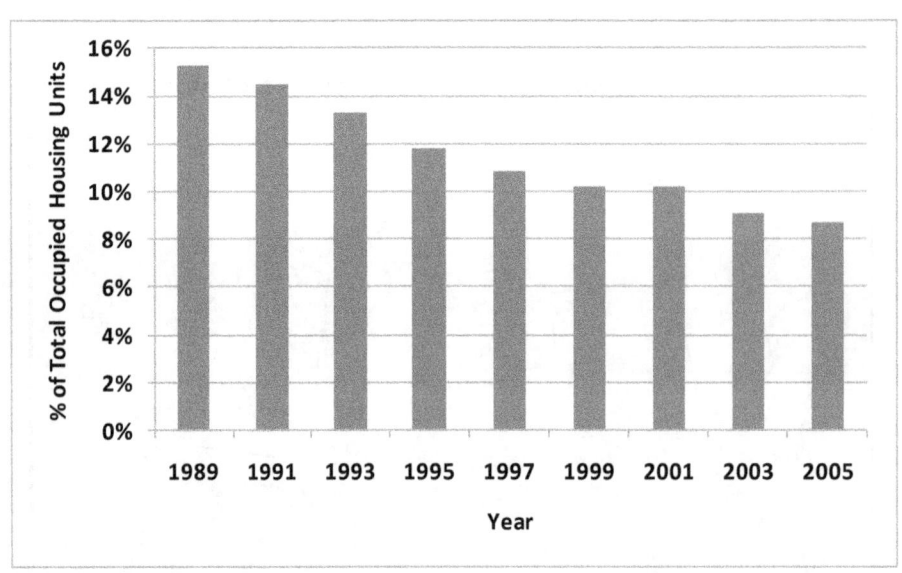

Figure 3-2.   Declining Trend in U.S. Housing Units Using Wood Fuel: 1989–2005

Source: U.S. Housing and Urban Development [HUD]. 2008. *American Housing Survey for the United States.* Multiple Years. Table 3-5. Available at http://www.census.gov/hhes/www/housing/ahs/nationaldata.html.

roughly 15% of all occupied housing units used wood fuel. The proportion of wood-fuel users has declined relatively steadily throughout the past 20 years. By 2005, fewer than 9% of the total 109 million occupied households in the United States used wood fuel for heating, cooking, or heating water.

The indoor fireplace market illustrates the continuing decline in wood fuel use over the past decade (HPBA, 2010a). As discussed in the next section, consumers are trending toward gas fireplaces instead of wood-fueled fireplaces. Fireplace manufacturers report that shipments of wood-fired factory-built fireplaces have been declining over the past decade as a result of the weakening new home construction market and the shift in consumer preferences toward gas fireplaces in the new homes that are being built (HPBA, 2010a). Of new home fireplaces, only 35% burn wood, whereas 65% are fueled by gas (HPBA, 2010a). It should be noted, however, that this trend has been arrested to some degree in recent times as the cost of wood fuel for heater/stove/furnace heating has come in line with the cost of oil and gas stove/heater/furnace heating, and trends show an increase in wood heating in households as shown in the unit cost memo prepared for this proposed rule.[8]

---

[8] U.S. EPA. Memorandum. Unit Cost Estimates of Residential Wood Heating Appliances. February 21, 2013. Prepared by EC/R, Inc.

3.2.4 Substitution Possibilities

The availability of close substitutes for wood heating equipment is largely contingent on two key factors: (1) the consumer's heating needs and preferences and (2) the price and availability of an alternative heating source. As discussed in Section 3.2.1, consumers tend to fall into one of three demand segments depending on their desired end use for their heating device. Each consumer group displays varying degrees of substitutability. The relative price of alternatives is also an important aspect of product substitution, which includes the cost of the heating equipment itself and the price and availability of the fuel it requires.

For most consumers looking for whole-house heat or single-room heat, gas or electric heat provides a common substitute for wood fuel. Electricity can power central heating systems for whole-house heat and smaller space heaters for single rooms. Since the majority of American households have easy access to electric power, these home heating options are often a convenient and low-cost alternative to wood heat. Gas-powered central furnaces and room heaters and oil-powered central heating systems are also on the market for residential use (DOE, 2009). Although most consumers have homes equipped for gas or electric power, more rural areas of the country have limited access to reliable utilities. In these regions, electric or gas heat may not be an available or cost-efficient choice relative to wood heat.

Recreational or aesthetic wood-fired appliances have fewer direct substitutes. Traditional indoor fireplaces and masonry fireplaces can be outfitted for burning natural gas rather than wood. Consumers may have a personal preference for one over the other. Wood fuel can be messy and somewhat difficult to store, whereas natural gas can be more convenient and easier to use, and gas furnaces can be much more efficient. Outdoor recreational appliances may be difficult to substitute directly because many consumers desire the aesthetic effect created by a wood-burning fire pit or chiminea. Outdoor charcoal or gas grills provide an alternative for outdoor wood-fired cooking appliances, but consumers may not consider these a direct substitute. It should be noted that outdoor recreational wood-burning appliances such as fire pits, chiminea, and grills are not covered in this proposal.

3.2.5 Price Elasticity of Demand

Price elasticity of demand is a concept in economic theory. It is a numeric measure of the sensitivity of demand following an increase or decrease in the product's price. The level of sensitivity is determined by a number of factors that include the availability and price of substitutes (e.g., other types of heating equipment, gas or electric space heaters and furnaces) and the price of complements (wood fuels).

In preparing this profile, we searched for, but were unable to identify, any empirical estimates of the price elasticity of demand for residential wood heating equipment in recent times. An estimate of -1.6 was derived for use in the RIA for the current Residential Wood Combustion NSPS (EPA, 1986). Although numerous articles estimate the elasticity of demand for residential energy and heating fuels, these estimates focus almost exclusively on electricity, natural gas, and fuel oil. These estimates find that residential energy and heating fuel demand is relatively inelastic (i.e., there are only very small changes in demand in response to an increase in energy or fuel prices). A recent RAND report suggests that in the short term, demand for electricity and natural gas in residential markets is relatively inelastic (Bernstein and Griffin, 2005). However, the authors of the report also note that sustained higher energy prices in the long term may result in demand for energy becoming more elastic as consumers have time to identify more energy-efficient options.

In the absence of empirical estimates, we offer a qualitative discussion of the key determinants of the price elasticity of demand to provide a general sense of whether consumer demand is elastic or inelastic. As mentioned earlier, the determinants of elasticity include the degree of substitutability, product necessity, and duration of the price increase.

There are a number of close substitutes for residential wood heating devices that include electric and gas furnaces and space heaters. The extent to which consumers are able to substitute between these options is likely to vary depending on geographic location. Overall, the presence of good substitutes will increase the elasticity of demand for wood heating equipment. In contrast, if locally-available alternative heating fuels (i.e., electricity, and fuel oil) are relatively higher priced, it may make switching away from wood heating equipment less likely and, ultimately, make demand for wood heating equipment inelastic. Also, the elasticity may depend on whether the fuel in question is a secondary source of fuel instead of a primary fuel source.

Finally, the magnitude of the cost for residential wood heating equipment may also increase the elasticity of demand. Consumer demand tends to be more elastic when the price of the good represents a large proportion of consumer income (Bernstein and Griffin, 2005). In other words, consumers become more sensitive to small price changes when considering the purchase of a large household appliance (e.g., refrigerator, oven range, or heating system).

3.3   Industry Organization

A review and description of market characteristics (i.e., geography, product differentiation, product transportation, entry barriers, and degree of concentration) can enhance our understanding of the mechanisms underlying the wood heating equipment industry. These characteristics provide indicators of a firm's ability to influence market prices by varying the quantity of product it sells. For example, in markets with large numbers of sellers and identical products, firms are unlikely to be able to influence market prices via their production decisions (i.e., they are "price takers" and operate in highly competitive markets). However, in markets with few firms, significant barriers to entry (e.g., licenses, legal restrictions, or high fixed costs), or products that are similar but can be differentiated, a firm may have some degree of market power (i.e., to set or significantly influence market prices). In addition, if a product is difficult to transport over long distance (due to weight or hazardous nature), then the market size may be more restricted than one might expect, all other things being equal.

3.3.1 *Market Structure*

Market structure characterizes the level of competition and determines the extent to which producers and sellers can influence market prices. Economic market structure typically focuses on the number of producers and consumers, the barriers to market entry, and product substitutability.

The residential wood heater market contains a number of large producers selling a number of differentiated products along with a large number of small producers. These characteristics suggest a quasi-monopolistic competitive market (i.e., somewhere between highly competitive and less competitive) for large producers who will have some influence over market prices. For small producers, the market will be highly competitive in nature. In addition, existing regulatory requirements for product testing and certifications represent a barrier to market entry for new producers of wood heating devices. Competition in this market may be further constrained by transportation costs due to the weight of these products. A similar assessment was determined in the 1986 study by the American Enterprise Institute (AEI) and Brookings Institution Joint Center for Regulatory Studies.

The AEI-Brookings report also identified several key factors that influence manufacturers' pricing decisions. These factors included production prices, prices of similar products sold by competitors, transportation costs, combustion technology and efficiency, and consumers' ability to differentiate products based on brand name and efficiency.

Price elasticity of supply is a numeric measure of the industry's response to a small percentage increase in the product price (Landsburg, 2005). The law of supply suggests producers supply greater quantities at higher prices as a result of increasing marginal returns for each additional unit produced as the average cost per unit of output declines. As a result, the elasticity of supply for most industries is positive. Determinants of supply elasticity are flexibility of sellers to adjust production and the time period being considered in estimating the elasticity. Most manufactured goods have an elastic supply, meaning that sellers can adjust production quickly in response to a change in prices (Mankiw, 1998). Industries with excess plant capacity are likely to have elastic supply as sellers can ramp up production in a relatively short time frame.

Based on 2006 plant capacity utilization data as shown in Figure 3-3, the heating equipment manufacturing industry averaged 60% utilization, growing from 59% in 2002 to a maximum utilization of 65% in 2005 and then falling to 54% in 2006 (U.S. Census Bureau, 2007). Similar statistics are not available for more recent years because this survey was discontinued after 2006. The available data suggest that there is ample existing capacity to increase production in the short and long terms, assuming an increase in price of residential wood-burning heating equipment.

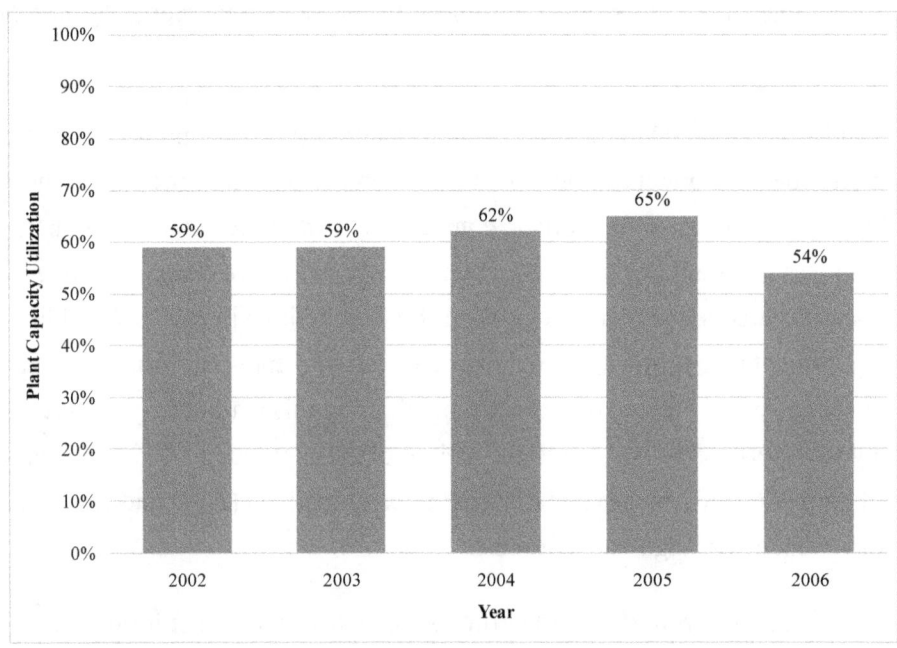

Figure 3-3.   Annual Plant Capacity Utilization for Heating Equipment Manufacturers (NAICS 333414): 2002–2006

Source: U.S. Census Bureau. 2007. *Survey of Plant Capacity: 2006.* "Table 1a. Full Capacity Utilization Rates by Industry Fourth Quarter 2002–2006." Census Bureau, Washington DC. Report No. MQ-C1(06).

3.3.2   Manufacturing Plants

Since 1988, the change in the number of residential wood-fired heater producers is unclear. The U.S. Economic Census reports that between 1992 and 2007 the number of establishments (places of business) in the industry has remained unchanged. Alternatively, the industry association (HPBA) has suggested that the number of manufacturers of wood-fired heaters fell by 80% following the 1988 NSPS, down from approximately 500 to roughly 120 manufacturers today (Houck and Tiegs, 2009). The difference between the 2 estimates is thought to be due to the large number of "backyard welders" in 1988 who built handmade stoves in their backyard as a sideline rather than their main source of income and chose not to attempt to develop competitive designs for the marketplace after the 1988 NSPS was promulgated.

For this analysis, we were able to identify 635 firms in the residential wood heating equipment industry in the United States. RTI developed this estimate leveraging a number of different sources that included EPA's official list of certified wood heater manufacturers, Dun & Bradstreet's online company database, and a number of industry association membership lists. The estimate includes the manufacturers listed on EPA's official certification lists (~120 manufacturers). We then expanded this list to include manufacturers of masonry heaters and outdoor wood boilers and manufacturers of non-heating devices, such as cook stoves, outdoor fireplaces, and bake ovens. Table 3-7 reports the count of U.S.-based companies in the industry by major business type.

Table 3-7. Number of U.S. Companies by Business Type

Business Type	Number of Companies	Reported Sales 2008 ($1,000s)	U.S. Market Share (% of Net Sales)
Manufacturers	577	$1,285,800	96.60%
Masonry contractors	24	$7,200	0.54%
Wholesalers, distributors	19	$34,200	2.57%
Residential construction	10	$3,200	0.24%
Retailers	5	$600	0.05%
U.S. Totals	**635**	**$1,331,000**	**100%**

Sources: Dun & Bradstreet *Marketplace*, a company database. RTI International calculations.

Residential wood heater manufacturers account for over 90% of the firms in the industry and span 14 different NAICS codes, of which 560 are categorized as NAICS 333414, as establishments primarily engaged in manufacturing heating equipment (except electric and warm air furnaces), such as heating boilers, heating stoves, floor and wall furnaces, and wall and baseboard heating units (Census Bureau, 2010f). An average manufacturer may produce anywhere from one to five technically different products (HPBA, 2010a). Manufacturers dominate the market, accounting for over 96% of sales for the industry in 2008.

Masonry contractors are the second largest group of businesses, accounting for 5% of the companies, and almost all masonry contractors are classified as NAICS 238140 establishments primarily engaged in masonry work, stone setting, brick laying, and other stone work for new construction, additions, alterations, maintenance, and repairs (U.S. Census Bureau, 2010f). Masonry contractors account for less than 1% of the industry sales. The remaining 5% of businesses are classified as residential construction contractors, wholesalers, distributors, and retailers. Residential construction contractors are primarily associated with design construction and installation of masonry heaters, outdoor fireplaces, and hydronic heaters. Companies classified as wholesalers, distributors, and retailers do not manufacture products but may be tied exclusively to a single brand or manufacturer, while others distribute and sell multiple products and brands.

3.3.3 Location

The industry is, for the most part, co-located in areas of the country with the largest demand for winter heating. Over 50% of U.S.-owned companies are located in 10 states in the northern half of the country. The largest number of companies is located in California, with additional concentrations in the Northwest, Northeast, the upper Midwest, and Central Plains. Table 3-8 reports the number of U.S. companies for the top 10 states. Additionally, approximately 104 foreign-based companies operate in the United States, two-thirds of which are Canadian-based companies.

Table 3-8. U.S. Wood Heat Equipment Industry by Geographic Location

Location	Business Count	% of Total U.S. Industry
California	63	10%
Pennsylvania	36	6%
Minnesota	35	6%
New York	33	5%
Washington	31	5%
Ohio	29	5%
Texas	29	5%
Wisconsin	26	4%
Michigan	25	4%
Illinois	21	3%
U.S. Total	**635**	**86%**
Canada	67	9%
Other foreign	37	5%
Industry Total	**739**	**100%**

Sources: Dun & Bradstreet *Marketplace*, a company database. RTI International calculations.

3.3.4 Company Sales and Employment

Overall sales for the residential wood heating industry totaled more than $1.3 billion in 2008. Based on company data obtained for this profile, the industry employs approximately 17,000 workers annually. Previous analysis suggests that the industry relies on seasonal labor, ramping up production in months leading up to winter and reducing employment and production during the warmer parts of the year (AEI, 1986). Table 3-9 presents median sales and employment for the industry by business type.

Table 3-9. U.S. Sales and Employment Statistics by Business Type

Business Type	Number of Companies	Median Sales 2010 ($1,000s)	Median Employment per Company
Manufacturers	577	$204	4
Masonry contractors	24	$102	3
Wholesalers, distributors	19	$510	5
Residential construction	10	$102	2
Retailers	5	$102	2
U.S. Totals	635	$204	4

Sources: Dun & Bradstreet *Marketplace*, a company database. RTI International calculations. Median sales estimates are escalated to 2010 from 2008 using the GDP implicit price deflator. The resulting escalation ratio for these years is 1.022.

Firms manufacturing heating equipment (except electric and warm air furnaces), such as heating boilers, heating stoves, floor and wall furnaces, and wall and baseboard heating units (NAICS 333414), are classified as small by the Small Business Administration (SBA, 2013) if they have fewer than 500 employees. Looking across the 14 manufacturing-related NAICS codes in our analysis, we find that approximately 90% of manufacturers are considered small businesses based on their reported employment compared with the SBA threshold. SBA classifies wholesalers and distributors as small if their employment is fewer than 100 workers. Approximately 68% of the industry's wholesalers and distributors are considered small based on the employment data obtained for this analysis.

SBA thresholds for masonry, construction, and retail firms are based on annual sales. SBA standards for NAICS codes under these business types range between $14 and $33 million in annual revenue. As reported in Table 3-9, median sales in these business categories are far below the range of SBA standards. As one would expect, our analysis finds that all 39 firms are considered small based on their reported annual sales compared with the SBA standards for their respective NAICS code classifications.

3.3.4.1 Profits of Affected Entities

Table 3-10 reports profit margins for manufacturers, masonry contractors, and wholesalers and distributors. The profit margin represents an average of reported profit per unit sales across the industry classified by the 6-digit NAICS code.

Table 3-10. Profit Margins for NAICS 333414, 238140, and 423720: 2008

NAICS Code	NAICS Description	Profit Margin	Industry Sales (10^6)
333414	Heating Equipment Manufacturers	4.3%	$70,965
238140	Masonry Contractors	4.7%	$9,676
423720	Plumbing and Heating Equipment Supplies (Hydronics) Merchant Wholesalers	3.4%	$58,907

Source: The Risk Management Association. 2008. *Annual Statement Studies, Financial Ratio Benchmarks 2008–2009.* Risk Management Association, Philadelphia: 2008.

3.4   Residential Wood Heater Market

Residential wood heating device shipments in the United States were relatively consistent from year to year between 1998 and 2005, according to the HPBA's reported hearth industry shipment data (2009). Since 2005, total industry shipments on average have declined annually by 24%. Industry experts attribute this decline in large part to the broader economic downturn and poor housing market. Renewable energy tax rebates offered in 2008 provided some relief for pellet-fueled devices, resulting in a 1-year increase in shipments of 161%, only to steeply decline again in 2009. This reflects the impact that the renewable energy tax rebates can have on wood burning appliances depending on the size and duration of the rebates. Table 3-11 presents shipment volumes by product type in 2008.

Outdoor wood boilers (or hydronic heaters) are a relatively new product in the market since 1990. Previous studies have reported annual growth in sales of between 30 and 128%, with over 155,000 outdoor wood boilers in use in the United States in 2006 (NESCAUM, 2006). Sales have been regionally focused in the Northeast (especially the Great Lakes region) and Midwestern states. The NESCAUM report predicted that over 500,000 outdoor wood boilers will be in use before the end of 2010 if trends in annual sales continue to follow growth rates observed between 1990 and 2006.

Market data for coal-burning stoves are very limited. However, anecdotal evidence suggests that coal stove use is limited to major coal states, including Pennsylvania, West Virginia, and Indiana, where coal is abundant and cheap relative to other heating fuels (Dagan,

Table 3-11. Unit Shipments and Percentage of Total Units by Product Type: 2008

Product Type	Units	% of Total Units
Wood stove	166,527	33%
Pellet stove	130,381	26%
Biomass stove	6,819	1%
Wood fireplaces[a]	180,966	36%
Outdoor fireplaces	6,302	1%
Masonry heaters	730	0%
Hydronic central heating systems	13,385	3%
Total	**505,110**	**100%**

[a] Wood fireplaces in this table include both factory-built and site-built models.

Source: Frost & Sullivan. 2010. *Market Research Report on North American Residential Wood Heaters, Fireplaces, and Hearth Heating Products Markets.* Prepared for EC/R Inc.

2005). Most of the major stove manufacturers feature at least one coal-burning stove model. However, at the time of writing this profile, we were unable to locate any reliable estimate of shipments in the United States for coal stoves.

3.4.1 Market Prices

Residential wood combustions device prices range from $200 to $50,000 depending on the product type and characteristics. Consumers who purchase these products must also consider the costs of installation, which range between $300 and $6,000 on average. Tables 3-12 and 3-13 report the average cost of installation and purchase price for residential wood combustion devices.

Table 3-12. Installation Costs for Average System by Product Type (North America): 2008

Product Type	Installation Cost
Wood stove	$500
Pellet stove	$300
Biomass stove	$300
Wood fireplaces	$600
Outdoor fireplaces	$350
Masonry heaters	$6,000
Hydronic central heating systems	$2,000

Source: Frost & Sullivan. 2010. *Market Research Report on North American Residential Wood Heaters, Fireplaces, and Hearth Heating Products Markets.* Prepared for EC/R Inc.

Table 3-13. Manufacturers' Price by Product Type (North America): 2008

Product Type	Average Price	Price Range
Wood stove	$848	$200 to $2,800
Pellet stove	$1,279	$300 to $3,500
Biomass stove	$1,403	$350 to $4,000
Wood fireplaces	$450	$150 to $5,000
Outdoor fireplaces	$755	$250 to $6,000
Masonry heaters	$9,041	$4,000 to $15,000
Hydronic central heating systems	$7,433	$5,000 to $35,000

Source: Frost & Sullivan. 2010. *Market Research Report on North American Residential Wood Heaters, Fireplaces, and Hearth Heating Products Markets,* Figure 2.6. Prepared for Ec/R Inc.

Given the specialized skills and materials required to construct a masonry heater, it is not surprising that this product has the highest average market price. Hydronic heaters are the second most expensive product partly because of the additional material requirements. The price of freestanding stoves and fireplace inserts varies depending on the fuel it burns. Biomass stoves are almost twice as expensive as cord wood-burning stoves because biomass stoves are more similar in construction to pellet stoves. Although no price data exist on coal-burning stoves, costs are comparable to traditional cord wood stoves. Coal stove prices for 2010 collected for this profile averaged $1,338 and ranged between $500 and $3,000 depending on the size and manufacturer.

3.4.2 *International Competition*

The U.S. market for wood-fueled heating products has been concentrated on the local scale in recent years. Manufacturers concentrate production where wood heat is in demand, which is in the Northeast and Northwest. Some regions of the country have specific emissions requirements on wood burning, so consumers may be restricted to buying stoves and heaters that can cater to local regulations (Frost & Sullivan, 2010). Domestic producers have traditionally faced some competition from European manufacturers in certain wood heat markets, but Asian manufacturers have been gaining market share, especially in the EPA-certified wood stove and currently exempt single-burn-rate stove markets (Frost & Sullivan, 2010).

Asian-based companies, especially those in China, have the advantage of relatively low overhead and labor costs compared with other companies worldwide (Frost & Sullivan, 2010). Although the products coming from these producers are lower in price, they are also lower in quality (Frost & Sullivan, 2010). However, money-conscious consumers have been willing to settle for lower quality stoves as the economy remains uncertain (Frost & Sullivan, 2010). Companies from all over the world have been moving some manufacturing operations to China in an attempt to compete with Asian producers through low-cost production (Frost & Sullivan, 2010). Still, U.S. manufacturers are likely to see increased competition from Asia in the future.

The masonry heater industry is one in which foreign manufacturers play a substantial role. Over two-thirds of masonry heaters installed in the United States are manufactured outside of the country, principally by one manufacturer (Seaton, 2010). Most United States companies build around 15 masonry heaters per year, typically constructed onsite by masons. Canadian and European producers sell masonry products through U.S. distributors, but most of these companies do not manufacture within the United States (Seaton, 2010). Some stove companies perform research and development, as well as assembly of wood stoves in the United States, but import cast parts and components from Europe and China (HPBA, 2010a). The pellet stove industry has seen increasing foreign competition in recent years. Many of the foreign manufacturers have made the business decision to sell products through American-owned businesses and thus the costs of EPA certification are sometimes passed on to the American seller/importer/licensee.

3.4.3   Future Market Trends

While there has been a steady decline in the residential markets for wood heaters, fireplaces, and hearth products, increases in oil and gas prices have led to substitution back to wood as a source of heat in 2007, in which a growth rate of 16.4% between 2007 and 2008 took place in these markets. However, demand for these products fell victim to the recession in 2009 (Frost & Sullivan, 2010). A weak residential construction market coupled with a tight credit market decreased overall demand in the market for wood heating products, leading analysts to project a 2009 growth rate of −36.1% (Frost & Sullivan, 2010). The growth forecast for 2010 is expected to improve relative to 2009 to a rate of −4.1%, due in part to the residual effects of the severe 2009 winter temperatures and the financial incentive provided by the federal energy efficient tax credit (Frost & Sullivan, 2010).

As the economy continues to recover beyond 2010, demand should trend upward as consumers look to cut heating costs with wood and biomass (Frost & Sullivan, 2010). New home construction and increased credit availability will further foster demand, which is expected to grow at a compound annual rate of 4.1% from 2009 to 2015 (Frost & Sullivan, 2010). The current regional demand patterns are expected to continue, with the Northeast and Northwest regions of the country driving wood fuel combustion demand, but analysts anticipate that the wood heat product market will be embraced in other areas of the country in which wood and biomass are viable and inexpensive fuel sources (Frost & Sullivan, 2010).

Although the overall residential wood heat market is expected to grow, there may be variation in demand between individual product segments. Pellet and biomass stoves are expected to lead the way in demand as consumers look for options with sustainable fuel sources and cleaner-burning technologies (Frost & Sullivan, 2010). Outdoor wood boilers (hydronic heaters) saw a surge in demand throughout the 1990s and mid-2000s, a trend that is projected to continue (Northeast States for Coordinated Air Use Management, 2006). Future demand for primary and secondary wood-burning heating devices will be somewhat dependent on the price of wood fuel relative to electric and gas heat, as well as consumer preferences. Since fireplaces and masonry fireplaces typically are not effective heaters and purchases are based on the aesthetic value rather than function, future demand will likely stay in line with consumer preferences.

SECTION 4
BASELINE EMISSIONS AND EMISSION REDUCTIONS

4.1   Introduction

This section presents the baseline emissions for the pollutants emitted by affected units and also the resulting emissions after imposition of the two options considered for the proposed NSPS. We present the baseline emissions and emission reductions for $PM_{2.5}$ and also for other pollutants from affected units such as volatile organic compounds (VOCs) and carbon monoxide (CO). Baseline emissions were calculated using a 2008 base inventory and were then projected to future years beyond the promulgation of the rule in 2014 to 2022 and beyond, using emissions factors specific to the category of the affected unit (e.g., certified wood stove, pellet stove). These emissions factors are listed in the emissions memorandum for this proposed rule (Ec/R, 2013). Emission reductions were calculated from the baseline emissions based on the considered emissions limits for each appliance type affected for each option analyzed, and the emission reductions were used as inputs to the benefits analysis presented in Section 7.

4.2   Background to Emissions Estimates

We used the EPA Residential Wood Combustion (RWC) emission estimation tool,[10] which is an Access[TM] database that compiles nationwide RWC emissions using county level, process specific data and calculations. We summed the nationwide number of appliances and total tons of wood burned for each of the relevant product categories in the inventory.

Table 4-1.   RWC Emission Inventory Categories Used

Woodstove: fireplace inserts; EPA certified; non-catalytic
Woodstove: fireplace inserts; EPA certified; catalytic
Woodstove: freestanding, EPA certified, non-catalytic
Woodstove: freestanding, EPA certified, catalytic
Woodstove: pellet-fired, general
Woodstove: freestanding, non-EPA certified
Hydronic heater: outdoor
Furnaces: indoor, cordwood

---

[10] rwc_2008_tToolv4.1_feb09_2010.zip.

We then made some adjustments/assumptions to the baseline RWC inventory. First, we deleted data in the RWC for non-certified stoves and inserts, as these cannot be sold. With the exception of wood stoves, we applied the $PM_{2.5}$ emission factors for each class to the total tons of wood burned and calculated an average emission rate/appliance/category. In the case of wood stoves, the RWC used an average of all PM_{10} AP-42 emission factors for wood stoves.[11] The RWC assumes that PM_{10} and $PM_{2.5}$ factors are identical. At a minimum, we believe that all new wood stoves meet the AP-42 PM_{10} emission factors for "Phase II" stoves (the current NSPS). As described below, we went a step further and assumed that all new shipments will meet the current Washington State limits, which are approximately 40% less than the current NSPS.

Second, we assumed that outdoor hydronic heaters and indoor hydronic heaters have the same emission profile.

Single burn rate stoves are not included in the RWC as separate identifiable units. We assumed that they would have the same baseline emission factor as freestanding non-certified woodstoves, i.e., 30.6 lb/ton of wood burned. We used the average tons burned per appliance factor as representative of these stoves as well.

Masonry heaters are not included in the RWC database, and we were unable to identify a surrogate emission factor which could be used to estimate tons/appliance emissions. Therefore, we were not able to estimate emissions from these appliances for the purpose of this analysis.

We used this subset of the RWC database to calculate a baseline average emission rate/appliance/category, including an adjustment of the RWC emission factor to the current Washington State limits where warranted. We multiplied the total tons of wood burned for the appliance by the RWC emission factor (adjusted as appropriate) to calculate the total tons of $PM_{2.5}$ emissions. We divided this value by the number of appliances in the category to calculate the baseline average $PM_{2.5}$ emissions per individual appliance, and these results are shown in Table 4-2.

---

[11] AP-42, Chapter 1.10, Residential Wood Stoves, Table 1.10-1. See: http://www.epa.gov/ttnchie1/ap42/ch01/final/c01s10.pdf.

Table 4-2. PM$_{2.5}$ Tons per Appliance Estimate (Baseline)

Emission Inventory Category	Pollutant	Baseline Emission factor (lb/ton)	Emissions (tons)	Tons per appliance/ yr
Woodstove: fireplace inserts; EPA certified; non-catalytic	Primary PM$_{2.5}$	8.76	5,371	0.0041
Woodstove: fireplace inserts; EPA certified; catalytic	Primary PM$_{2.5}$	9.72	2,023	0.0047
Woodstove: freestanding, EPA certified, non-catalytic	Primary PM$_{2.5}$	8.76	6,745	0.0077
Woodstove: freestanding, EPA certified, catalytic	Primary PM$_{2.5}$	9.72	3,769	0.0101
Woodstove: pellet-fired, general	Primary PM$_{2.5}$	3.06	1,798	0.0021
Hydronic heater: outdoor/indoor	Primary PM$_{2.5}$	27.6	50,427	0.1383
Woodstove: freestanding, non-EPA certified[a]	Primary PM$_{2.5}$	30.6	71,424	0.0324
Furnace: indoor, cordwood	Primary PM$_{2.5}$	27.6	2,471	0.0582
	Primary PM$_{2.5}$	30.6		

[a] Non-EPA certified wood stove emission factor and tons/appliance were used to represent single burn rate stoves.

4.2.1 Emissions Factors

The next step in the analysis was to develop emission factors representing potential NSPS options to reduce emissions. The following is a summary of the NSPS options considered in the proposal for each appliance type and examined in detail in Section 2. The NSPS options examined in this analysis ("Proposal" and "Alternative") are based on phased-in compliance dates, or "steps," for subcategories of appliances. Proposed Subpart AAA will regulate "room heaters" and includes adjustable burn rate stoves, single burn rate stoves, and pellet stoves. Proposed Subpart QQQQ will regulate "central heaters" and includes outdoor and indoor hydronic heaters and forced air furnaces. Proposed Subpart RRRR will regulate masonry heaters. Following is a summary of the current NSPS implementation assumptions for appliances within the subcategories under both the Proposed and the Alternative options. As mentioned in Section 2, the Proposed option is a 2-step standard with compliance dates of effective date and 5 years after the effective date for different appliances. For the purposes of our Proposed option analyses, we used 2014 and 2019, respectively. The Alternative option is a 3-step standard with compliance dates of effective date, 3 years after the effective date, and 8 years after the effective date. For the purposes of our Alternative option analyses, we used 2014, 2017, and 2022, respectively.

Subpart AAA ("room heaters"):

- These are adjustable burn rate, single burn rate, and pellet stoves: **Proposal**: Step 1 limit of 4.5 g/hr upon promulgation in 2014; and Step 2 limit of 1.3 g/hr five years after promulgation in 2019. **Alternative**: Step 1 limit of 4.5 g/hr upon promulgation in 2014; Step 2 limit of 2.5 g/hr three years after promulgation in 2017; and Step 3 limit of 1.3 g/hr eight years after promulgation in 2022. **Note**: The Step 1 limit is the 1995 Washington State standard for non-catalytic stoves; the Alternative Step 2 limit is the 1995 Washington State standard for catalytic stoves; and the proposed Step 2 (Alternative Step 3) limit is already met by the top performing catalytic, non-catalytic and pellet stove models, according to industry data.[12] Although previously unregulated and a less developed technology than adjustable burn rate stoves, single burn rate stove designs have been undergoing R&D in anticipation of the proposed NSPS and cleaner designs are nearly market-ready.[13]

Subpart QQQQ ("central heaters"):

- These are hydronic heaters (both outdoor and indoor): **Proposal**: Step 1 limit of 0.32 lb/mm BTU heat output upon promulgation in 2014; and Step 2 limit of 0.06 lb/mm BTU heat output five years after promulgation in 2019. **Alternative**: Step 1 limit of 0.32 lb/mm BTU heat output upon promulgation in 2014; Step 2 limit of 0.15 lb/mm BTU heat output three years after promulgation in 2017; and Step 3 limit of 0.06 lb/mm BTU heat output eight years after promulgation in 2022. **Note**: The Step 1 limit is the EPA "Phase 2 " voluntary program limit already met by 36 hydronic heater models (27 cord wood and 9 pellet models) built by 17 U.S. manufacturers; the Alternative Step 2 limit is already met by 11 hydronic heater models (6 cord wood and 5 pellet models) built by 6 U.S. manufacturers; and the proposed Step 2 (Alternative Step 3) limit is already met by 4 hydronic heater models (2 cord wood and 2 pellet models) built by 2 U.S. manufacturers[14], as well as over 100 European manufacturers per test method EN 303-05.[15]

- Forced Air Furnaces: **Proposal**: Step 1 limit of 0.93 lb/mm BTU heat output upon promulgation in 2014; and Step 2 limit of 0.06 lb/mm BTU heat output five years after promulgation in 2019. **Alternative**: Step 1 limit of 0.93 lb/mm BTU heat output upon promulgation in 2014; Step 2 limit of 0.15 lb/mm BTU heat output three years after promulgation in 2017; and Step 3 limit of 0.06 lb/mm BTU heat output eight years after promulgation in 2022. **Note**: The Step 1 limit is based on test data from

---

[12] Final HPBA Heater Database version 2/25/10, EC/R received from Bob Ferguson for HPBA on 4/26/10
[13] 2/8/13 telephone discussion between Gil Wood, USEPA, and a manufacturer of single burn rate stoves.
[14] See list of cleaner hydronic heaters participating in EPA's voluntary program at http://www.epa.gov/burnwise/owhhlist.html
[15] European Wood-Heating Technology Survey: An Overview of Combustion Principles and the Energy and Emissions Performance Characteristics of Commercially Available Systems in Austria, Germany, Denmark, Norway, and Sweden; Final Report; Prepared for the New York State Energy Research and Development Authority; NYSERDA Report 10–01; April 2010.

development of Canadian standard B415.1-10[16] and conversation with industry regarding cleaner forced air furnace models currently being tested in R&D[17]. Forced air furnace designs able to meet the Alternative Step 2 and proposed Step 2 (Alternative Step 3) limits may be based on technology transferred from hydronic heater designs.

Subpart RRRR including masonry heaters:

- Masonry Heaters: **Proposal / Alternative** (same): Step 1 limit of 0.32 lb/mm BTU heat output upon promulgation in 2014 for large manufacturers (defined as manufacturers constructing ≥ 15 masonry heaters per year), with a 5-year (2019) small volume manufacturer compliance extension (for companies constructing < 15 units/year). No other phased-in limits are being proposed. **Note**: Based on data submitted by the Masonry Heater Association[18], over 10 models already meet this limit.

We developed adjusted emission factors to reflect the NSPS options discussed above, which were then used to calculate new average tons of emissions per appliance for each RWC appliance type. Reasonable adjustments were assumed for NSPS emission factors (as noted below) in order to not overstate emission reductions under the NSPS options; actual emission reductions may be somewhat greater than reductions resulting from our emission factor adjustments for the purpose of this analysis. Following is a description of how the RWC factors were adjusted:

- Woodstove: all EPA certified. We determined the ratio of emissions between the existing NSPS limits compared to the Washington state standards, for they are tighter than the existing NSPS and have been in existence since 1995. For both catalytic and non-catalytic devices, the Washington standard is 60% of the NSPS. We assumed this same ratio would apply to the emissions factors and multiplied the RWC emission factor by 60%. We used these adjusted RWC emission factors (shown in Table 4-2) as both baseline and Step 1 emission factors for catalytic and non-catalytic stoves. We made the reasonable assumption (in terms of estimating potential emission reductions) that the Step 1 emission factor was the same as the baseline emission factor, because nearly all current wood stove models already meet the Step 1 limit according to industry data.[19] We also reasonably assumed that the Alternative Step 2 emission factor for catalytic stoves was the same as the baseline and Step 1 emission factor because approximately 90% of the current catalytic models already meet the Alternative Step 2 limit, according to industry data.[10] For the Alternative Step 2

---

[16] CSA B415.1-10, Performance Testing of Solid-Fuel-Burning Heating Appliances. Appendix D. March 2010.

[17] 2/8/13 telephone discussion between Gil Wood, USEPA, and a manufacturer of forced air furnaces.

[18] Attachment to 3/25/2011 e-mail from Timothy Seaton of Timely Construction to Gil Wood and Mike Toney of USEPA.

[19] Final HPBA Heater Database version 2/25/10, EC/R received from Bob Ferguson for HPBA on 4/26/10.

emission factor for non-catalytic models, we scaled the Step 1 emission factor by the ratio of the Alternative Step 2 standard to the Step 1 standard (or 2.5/4.5 = 0.55). Likewise, for the proposed Step 2 (Alternative Step 3) emission factor for both catalytic and non-catalytic models, we scaled the Alternative Step 2 emission factor by the ratio of the proposed Step 2 (Alternative Step 3) standard to the Alternative Step 2 standard (or 1.3/2.5 = 0.52). For consistency with our shipment data and because the RWC database provides four separate emission factors for catalytic and non-catalytic, freestanding models and fireplace inserts, we used the weighted average value for all four wood stove types to represent the total population of adjustable burn rate woodstoves. Finally, we multiplied the resulting emission factors by the total tons burned for the appliance type (provided by the RWC database) and then divided that by the appliance population (also provided by the RWC database) to derive the tons/appliance of $PM_{2.5}$ emissions. The emission factors and tons/appliance are shown in the green rows in Table 4-3.

- Woodstove: pellet fired, general. We used the RWC emission factor shown in Table 4-2 as both the baseline and Step 1 emission factor for pellet stoves because nearly all current pellet stove models already meet the Step 1 standard according to industry data.[20] The RWC emission factor for pellet stoves is quite low compared to other appliances, which leaves little room to adjust the factor. For the Alternative Step 2 emission factor, we reduced the Baseline/Step 1 emission factor by only 10%, an adjustment based on industry data[11] that most pellet models (80%) already meet the Alternative Step 2 level. For the proposed Step 2 (Alternative Step 3) emission factor, we scaled the Alternative Step 2 emission factor by the ratio of the proposed Step 2 (Alternative Step 3) standard to the Alternative Step 2 standard (or 1.3/2.5 = 0.52). We multiplied the resulting emission factors by the total tons burned for pellet stoves and then divided that by the pellet stove appliance population to derive the tons/appliance of $PM_{2.5}$ emissions. The emission factors and tons/appliance are shown in the orange row in Table 4-3.

- Woodstove: freestanding, non-EPA certified (single burn rate stoves). As described above, we assumed that the freestanding non-EPA certified woodstove emission inventory category includes the population of single burn rate stoves. We therefore used the RWC emission factor for freestanding non-EPA certified woodstoves (30.6 lb/ton) as the baseline emission factor for single burn rate stoves. For the Step 1 emission factor, we used the same emission factor as a certified non-catalytic stove meeting the Washington state standards (i.e., 8.76 lb/ton) because the same standard is being proposed for single burn rate stoves as for adjustable burn rate stoves. Likewise, we used the same emission factors used for non-catalytic stoves for the Alternative Step 2 and proposed Step 2 (Alternative Step 3) emission factors. We multiplied the resulting emission factors by the total tons burned for this appliance category and then divided that by the appliance population to derive the tons/appliance of $PM_{2.5}$ emissions. The emission factors and tons/appliance are shown in the grey row in Table 4-3.

---

[20] Final HPBA Heater Database version 2/25/10, EC/R received from Bob Ferguson for HPBA on 4/26/10.

4.2.1.1 Hydronic Heater: Outdoor/Indoor.

As noted above, we assumed that indoor hydronic heaters (a minority of the hydronic heater population) have the same emission profile as the outdoor hydronic heater appliance category provided in the RWC. According to the EPA voluntary hydronic heater program, the "phase 2" heaters that are presumed to represent the Level I NSPS option are 90% cleaner than older unqualified units.[21] We assume as described in our emissions memorandum that the majority of the existing inventory is represented by these unqualified units, and applied a 90% reduction to the RWC baseline emission factor shown in Table 4-2 (27.6 lb/ton) to derive the Step 1 emission factor (2.76 lb/ton). For the Alternative Step 2 emission factor, we assumed a 50% reduction in the Step 1 emission factor since the Alternative Step 2 limit is approximately 50% of the Step 1 limit. Likewise, for the proposed Step 2 (Alternative Step 3) emission factor, we scaled the Alternative Step 2 emission factor by the ratio of the proposed Step 2 (Alternative Step 3) standard to the Alternative Step 2 standard (or 0.06/0.15 = 0.40). We multiplied the resulting emission factors by the total tons burned for the hydronic heater RWC appliance category and then divided that by the hydronic heater appliance population to derive the tons/appliance of $PM_{2.5}$ emissions. The emission factors and tons/appliance are shown in the blue row in Table 4-3.

4.2.1.2 Furnace: Indoor, Cordwood

We used the RWC emission factor shown in Table 4-2 (27.6 lb/ton) as the baseline emission factor. For Step 1, we scaled the baseline emission factor by 75% (to 6.9 lb/ton) because background material provided in the CSA standards review process stated that the emission limit associated with this method would result in an approximately 75% reduction in emissions compared to a non-qualifying furnace.[22] The Alternative Step 2 and proposed Step 2 (Alternative Step 3) limits of 0.15 lb/mmBTU and 0.06 lb/mmBTU, respectively, are the same as the hydronic heater limits. The baseline emission factors for each appliance category are also the same. Therefore we used the same Alternative Step 2 and proposed Step 2 (Alternative Step 3) emission factors used for hydronic heaters (1.38 lb/ton and 0.55 lb/ton, respectively). We multiplied the emission factors by the total tons burned for the cordwood furnace RWC appliance category and then divided that by the furnace appliance population to derive the tons/appliance of $PM_{2.5}$ emissions. The emission factors and tons/appliance are shown in the lavender row in Table 4-3. See Table 4-3 for a summary of the emission factors and resulting

---

[21] See the EPA Burnwise Website: http://www.epa.gov/burnwise/participation.html.
[22] Review draft of CSA B415.1-10, Performance Testing of Solid-Fuel-Burning Heating Appliances. Appendix C. March 2010.

tons/appliance values for the baseline and NSPS options analyzed. Table 4-3 presents the baseline, Step 1, Alternative Step 2, and proposed Step 2 (Alternative Step 3) emission factors for each appliance type resulting from our assumptions and adjustments described above. We used the appropriate tons/appliance with annual shipment data to estimate annual $PM_{2.5}$ emissions based on the Proposed and Alternative phased-in implementation dates.

Table 4-3. NSPS Adjusted Factors for PM$_{2.5}$

Emission Inventory Category	Baseline Emission Factor (lb/ton)	Baseline Emissions/ Appliance (tons)	Tons/ Appl/yr	Step 1 Emission Factor	Tons/ Appl/yr	Alt. Step 2 Emission Factor lb/ton	Alt. Step 2 Tons/appl	Step 2 (Alt. Step 3) Emission Factor lb/ton	Step 2 (Alt. Step 3) Tons/appl
Woodstove: fireplace inserts; EPA certified; non-catalytic	8.76	5,371	0.0041	8.76	0.0041	4.82	0.0023	2.51	0.0012
Woodstove: fireplace inserts; EPA certified; catalytic	9.72	2,023	0.0047	9.72	0.0047	9.72	0.0047	5.05	0.0025
Woodstove: freestanding, EPA certified, noncatalytic	8.76	6,745	0.0077	8.76	0.0077	4.82	0.0042	2.51	0.0022
Woodstove: freestanding, EPA certified, catalytic	9.72	3,769	0.0101	9.72	0.0101	9.72	0.0101	5.05	0.0052
Woodstove: pellet-fired, general	3.06	1,798	0.0021	3.06	0.0021	2.75	0.0019	1.43	0.0010
Hydronic heater: outdoor	27.6	5,043	0.1383	2.76	0.0138	1.38	0.0069	0.55	0.0028
Furnace: indoor, cordwood	27.6	9,053	0.1032	6.9	0.0258	1.38	0.0052	0.55	0.0021
Single Burn Rate Stoves (freestanding, non-EPA certified)	30.6	20,447	0.0324	8.76	0.0093	4.82	0.0051	2.51	0.0027

4.2.2 Voluntary Programs

Within these emissions projections are the effects not only of rules but also of various voluntary programs managed by EPA and states. Studies have shown that fine particle ($PM_{2.5}$) concentrations in proximity to a typical outdoor wood boiler are likely to exceed the 24-hour National Ambient Air Quality Standards (NAAQS).[18] Thus, the EPA developed a hydronic heater voluntary program to encourage manufacturers to reduce impacts on air quality through developing and distributing cleaner, more efficient hydronic heaters. We developed the voluntary program because it could bring cleaner models to market faster than the traditional federal regulatory process. Phase 1[19] emission level (0.60 pounds per million British Thermal Unit (lb/MMBTU) heat input) qualifying[20] units are approximately 70% cleaner than typical unqualified units. After March 31, 2010, units that only meet the Phase 1 emission level are no longer considered "qualified models" under the voluntary program. Phase 2 emission level (0.32 lb/MMBTU heat output) qualifying units are approximately 90% cleaner than typical unqualified units. Typically, qualified models have improved insulation, secondary combustion, separation of the firebox from the water jacket, and the addition of improved heat exchangers.

In addition to the voluntary program, the EPA provided technical and financial support for the Northeast States for Coordinated Air Use Management (NESCAUM) to develop a model rule which several states have adopted to regulate hydronic heaters. The model rule is a starting point for local regulatory authorities to consider, and they may need to also adopt additional actions due to site-specific concerns, e.g., local terrain, meteorology, proximity of neighbors and other exposed individuals. Thus, some regulatory authorities have instituted additional requirements, including bans on hydronic heaters in some areas.

The EPA also developed a similar voluntary partnership program for low-mass fireplaces (engineered, pre-fabricated fireplaces) and site-built masonry fireplaces. The original partnership agreements were dated February 19, 2009, and pertained to low-mass fireplaces. On July 4, 2009, the program was expanded to other fireplaces, e.g., masonry fireplaces. Under this

---

[18] For more information on wood smoke health effects, See: "Smoke Gets in Your Lungs: Outdoor Wood Boilers in New York State," prepared by Judith Schrieber, Ph.D., et al., for the Office of the Attorney General of New York. August 2005. See also: "Assessment of Outdoor Wood-fired Boilers," prepared by NESCAUM, March 2006 (revised June 2006).

[19] Phase 1" and "Phase 2" emission levels refer to levels established in EPA voluntary programs. The earlier use of the term "Phase II" (with a Roman numeral) standard refers to standards established in the current subpart AAA for residential wood heaters.

[20] The terms "qualified" and "unqualified," or other similar terms, refer to models that meet the voluntary program performance levels. Later use of the terms "certified" and "uncertified," or other similar terms, refers to models that are deemed to be in compliance with the NSPS emission limits.

program, cleaner burning fireplaces are ones that qualify for the Phase 1 emissions level of 7.3 g of particles emitted per kg of fuel burned (approximately 57% cleaner than unqualified models) or the Phase 2 emissions level of 5.1 g/kg (approximately 70% cleaner than unqualified models). So far, 11 models have qualified under this voluntary program at the Phase 2 level. Typically, qualified models have improved insulation and added secondary combustion and/or a catalyst to reduce emissions. Some manufacturers have added closed doors to reduce the excess air and thus improve combustion. Some state and local agencies have needed to reduce emissions further and thus some have no-burn days and some have adopted bans of new fireplaces in some areas in order to attain the $PM_{2.5}$ NAAQS.

4.2.3 *Shipment Data Used to Estimate Baseline Emissions*

We used data in the Frost & Sullivan Market (F&S) report[21] on 2008 shipments by product category, and F&S revenue forecasts which incorporated the weak economy in years 2009 and 2010, to calculate the reduced number of shipments in years 2009 and 2010. Forced air furnaces were outside the scope of the F&S report. Instead, we used manufacturer estimates of total industry sales in 2008[22] and applied the F&S market factors to estimate shipments through 2010. The F&S wood stove numbers included both certified and non-certified stoves, so we estimated numbers of non-certified stove shipments out of the total reported wood stove category. [23] These shipments were deleted from the total wood stove category shipments. We expanded the 2008 single burn rate estimate using the F&S factors. Our estimates of annual shipments, truncated to 2022, are shown in Table 4-4. The full set of annual shipments data can be found in the emissions memo for this proposal.

For years 2011 through 2038 (for the proposed NSPS) and 2011 through 2041 (for the alternative approach) estimated shipments are based on a forecasted revenue growth rate of 2.0%, in keeping with the average annual growth in real GDP predicted by the Conference Board.[24] There is not a perfect correlation between shipments and revenue (for example, because of their higher unit cost, pellet stoves generate more absolute revenue than wood stoves), but as stated in our emissions memorandum, we think the overall trend in the projection is reasonable in the absence of specific shipment projections.

---

[21] Market Research and Report on North American Residential Wood Heaters, Fireplaces, and Hearth Heating Products Market. Prepared by Frost & Sullivan. April 26, 2010. pp. 31-32.

[22] NSPS Review and Comments. Confidential Business Information submitted by manufacturer. September 2010.

[23] Memo to Gil Wood, USEPA, from EC/R, Inc. Draft Estimated Emissions from Wood Heaters. February 15, 2013.

[24] 2013 Global Outlook projections prepared by the Conference Board in November 2012; http://www.conference-board.org/data/globaloutlook.cfm

Table 4-4.   Estimated Annual Shipments by Category, 2008-2022

	2008	2009	2010	2011	2012	2013	2014	2015	2016	2017	2018	2019
Wood stoves	126,527	80,851	77,617	79,169	80,752	82,367	84,015	85,695	87,409	89,157	90,940	92,759
Single burn rate stoves	40,000	25,560	24,538	25,028	25,529	26,039	26,560	27,091	27.633	28,186	28,750	29,325
Pellet stoves	130,381	83,313	79,981	81,581	83,212	84,876	86,574	88,305	90,072	91,873	93,710	95,585
Furnace: indoor, cordwood	41,000	26,199	25,151	25,654	26,167	26,690	27,224	27,769	28,324	28,891	29,468	30,058
Hydronic heating systems	13,385	8,553	8,211	8,375	8,543	8,713	8,888	9,065	9,247	9,432	9,620	9,813

Our cost effectiveness analysis (CE)[25] assumes a 20-year model design lifespan as well as a 20-year use/emitting appliance lifespan. These assumptions were made to best characterize the actual model design and use lifespans given that many models developed for the 1988 NSPS are still being sold (after 25 years), many "new" models still have the same internal working parts with merely exterior cosmetic changes, and most stoves in consumer homes emit for at least 20 years and often much longer. Therefore our CE analysis tracks shipments through year 2038 for the proposed NSPS and through year 2041 for the alternative approach (i.e., assuming a 20 year design life for a model meeting the Step 2 limit in year 2019 under the proposed NSPS, and for a model meeting the Alternative Step 3 limit in year 2022 under the alternative approach). A truncated summary of our actual shipment data which extended through years 2038 for the proposed NSPS and 2041 for the alternative approach can be found in the emissions memo for this proposal. See the CE analysis spreadsheets that are in the public docket for the complete shipment data.

4.3   Estimated PM$_{2.5}$ Emissions from Shipments of New Appliances

As described above, we calculated the average emissions per appliance type using the emission factor for each category multiplied by the inventory value of total tons of wood burned

---

[25] See cost effective (CE) spreadsheets including for PM$_{2.5}$ the *2019 Step 2 Wood Heater NSPS PM25 CE 7% Feb14_2013.xls* for the CE analysis supporting the proposed NSPS, and the *3 Step Wood Heater NSPS PM25 CE 7% Feb14_2013.xls* for the alternative approach. All of these spreadsheets are found in the public docket for this rulemaking.

divided by the number of appliances in the inventory population. This value was then multiplied by the number of shipments to calculate total emissions from each category per year under baseline conditions (i.e., in the absence of an NSPS). More information on these calculations is available in the emissions memorandum in the docket for this rulemaking.[26]

Table 4-5 on the next page shows a truncated summary of the estimated $PM_{2.5}$ emissions (in tons) under baseline conditions through year 2022. We then estimated emissions under the proposed NSPS (Table 4-6) and under the alternative approach (Table 4-7) based on the respective assumptions and phase-in timelines for each. Under the proposed NSPS, the Step 1 limit becomes effective in 2014 and the Step 2 limit in 2019; while under the alternative approach, the Step 1 limit becomes effective in 2014, the Step 2 limit in 2017, and the Step 3 limit in 2022. (Note that the proposed Step 2 limit is the same as the alternative Step 3 limit, although the compliance dates differ.) The emission estimates assume that the total number of shipped units meet the standard in the year the standard is implemented.

Tables 4-5, 4-6, and 4-7 show emission estimates out to year 2022 for comparison. These are truncated summaries. Our CE analysis tracks emission reductions out through 2057 for the proposed NSPS and through 2060 for the alternative approach, assuming a 20 year design life for a model meeting each phased-in limit, and best assumption that stoves shipped in the 20th year will be emitting in homes for another 20 years. See the cost-effectiveness analysis spreadsheets[27] for all years of emission data, both baseline and under each NSPS option considered. These spreadsheets are available in the public docket for this rulemaking.

Note: No emission tables are provided for Subpart RRRR regulating masonry heaters because emission estimates are not available for these appliances as explained earlier in this RIA section.

---

[26] Memo to Gil Wood, USEPA, from EC/R, Inc. Draft Estimated Emissions from Wood Heaters. February 15, 2013.
[27] See cost effective (CE) spreadsheets including for $PM_{2.5}$ the *2019 Step 2 Wood Heater NSPS PM25 CE 7% Feb14_2013.xls* for the CE analysis supporting the proposed NSPS, and the *3 Step Wood Heater NSPS PM25 CE 7% Feb14_2013.xls* for the Alternative approach.

Table 4-5. Estimated PM$_{2.5}$ Emissions (Tons): Baseline

	2008	2009	2010	2011	2012	2013	2014	2015	2016	2017	2018	2019	2020	2021	2022
Wood stoves	761	486	467	476	486	495	505	515	526	536	547	558	569	580	592
Single burn rate stoves	1,295	827	794	810	826	843	860	877	895	912	931	949	968	988	1,007
Pellet stoves	277	177	170	173	177	180	184	187	191	195	199	203	207	211	215
Furnace: indoor, cordwood	4,230	2,703	2,595	2,647	2,699	2,753	2,809	2,865	2,922	2,980	3,040	3,101	3,163	3,226	3,291
Hydronic heating systems	1,851	1,183	1,136	1,158	1,182	1,205	1,229	1,254	1,279	1,305	1,331	1,357	1,384	1,412	1,440
Total	8,414	5,376	5,161	5,265	5,370	5,477	5,587	5,699	5,812	5,929	6,047	6,168	6,292	6,417	6,546

Table 4-6. Estimated PM$_{2.5}$ Emissions (Tons): Proposed Option

	2008	2009	2010	2011	2012	2013	2014	2015	2016	2017	2018	2019	2020	2021	2022
Wood stoves	761	486	467	476	486	495	505	515	526	536	547	202	206	210	214
Single burn rate stoves	1,295	827	794	810	826	843	246	251	256	261	266	78	79	81	82
Pellet stoves	277	177	170	173	177	180	184	187	191	195	199	95	97	99	101
Furnace: indoor, cordwood	4,230	2,703	2,595	2,647	2,699	2,753	702	716	731	745	760	62	63	65	66
Hydronic heating systems	1,851	1,183	1,136	1,158	1,182	1,205	123	125	128	130	133	27	28	28	29
Total	8,414	5,376	5,161	5,265	5,370	5,477	1,760	1,795	1,831	1,868	1,905	464	473	482	492

Table 4-7.   Estimated PM$_{2.5}$ Emissions (Tons): Alternative Option

	2008	2009	2010	2011	2012	2013	2014	2015	2016	2017	2018	2019	2020	2021	2022
Wood stoves	761	486	467	476	486	495	505	515	526	373	380	388	396	404	214
Single burn rate stoves	1,295	827	794	810	826	843	246	251	256	144	147	149	152	155	82
Pellet stoves	277	177	170	173	177	180	184	187	191	176	179	183	186	190	101
Furnace: indoor, cordwood	4,230	2,703	2,595	2,647	2,699	2,753	702	716	731	149	152	155	158	161	66
Hydronic heating systems	1,851	1,183	1,136	1,158	1,182	1,205	123	125	128	65	67	68	69	71	29
Total	**8,414**	**5,376**	**5,161**	**5,265**	**5,370**	**5,477**	**1,760**	**1,795**	**1,831**	**906**	**925**	**943**	**962**	**981**	**492**

From the tables above, we see that the PM$_{2.5}$ emission reductions in 2022 are the same at 6,054 tons. The average of the annual emission reductions between the year of rule promulgation, 2014, and the year that both options are fully implemented (2022) is 4,825 tons for the Proposed option and 4,878 tons for the Alternative option.

4.4   Methodology for Estimating VOC Emissions from New Units

We used the same methodology described in Section 4.3 to develop emission estimates for VOC emissions. Using the RWC database, we developed an estimate of VOC emissions per appliance using baseline emission factors. Then, using the same NSPS phase-in assumptions and anticipated emission reductions (i.e., that VOC reductions are comparable to PM$_{2.5}$ reductions), we developed emission factors to be used in analyzing the NSPS options. Table 4-8 provides the VOC emission factors.

Table 4-8. NSPS VOC Emission Factors

Emission Inventory Category	Baseline Emission Factor (lb/ton)	Emissions (tons)	Tons/Appl/Yr	Step 1 Emission Factor (lb/ton)	Step 1 Tons/Appl/Yr	Alt. Step 2 Emission Factor (lb/ton)	Alt. Step 2 Tons/Appl/Yr	Step 2 (Alt. Step 3) Emission Factor (lb/ton)	Step 2 (Alt. Step 3) Tons/Appl/Yr
Woodstove: fireplace inserts; EPA certified; non-catalytic	12	7,357	0.0056	12	0.0056	6.6	0.0031	3.432	0.0016
Woodstove: fireplace inserts; EPA certified; catalytic	15	3,121	0.0073	15	0.0073	15	0.0073	7.800	0.0038
Woodstove: freestanding, EPA certified, non-catalytic	12	9,240	0.0106	12	0.0106	6.6	0.0058	3.432	0.0030
Woodstove: freestanding, EPA certified, catalytic	15	5,817	0.0155	15	0.0155	15	0.0155	7.800	0.0081
Woodstove: pellet-fired, general	0.041	24	0.00030	0.041	0.00003	0.037	0.00003	0.019	0.00001
Hydronic heater: outdoor	11.7	2,138	0.0586	1.17	0.0059	0.59	0.0028	0.234	0.0012
Furnace: indoor, cordwood	11.7	3,838	0.0437	2.925	0.0109	0.59	0.0022	0.234	0.0009
Single burn rate stoves (freestanding, non-EPA certified)	53	21,288	0.0561	12	0.0127	6.6	0.0070	3.432	0.0036

Using the same assumptions as we used for PM$_{2.5}$, we calculated VOC emissions at baseline and under each NSPS option based on a 20-year model design lifespan for appliance shipments as well as a 20-year appliance life. Tables 4-9 through 4-11 provide the time series of VOC annual emissions estimates between 2008 and 2022 for the baseline, and the NSPS options considered under the proposed NSPS revision.

Table 4-9. Estimated VOC Emissions (Tons): Baseline

	2008	2009	2010	2011	2012	2013	2014	2015	2016	2017	2018	2019	2020	2021	2022
Wood stoves	1,085	693	666	679	692	706	720	735	750	765	780	795	811	828	844
Single burn rate stoves	2,243	1,433	1,376	1,403	1,431	1,460	1,489	1,519	1,549	1,580	1,612	1,644	1,677	1,711	1,745
Pellet stoves	4	2	2	2	2	2	2	3	3	3	3	3	3	3	3
Furnace: indoor, cordwood	1,793	1,146	1,100	1,122	1,144	1,167	1,191	1,214	1,239	1,263	1,289	1,315	1,341	1,368	1,395
Hydronic heating systems	785	502	481	491	501	511	521	532	542	553	564	575	587	599	611
Total	5,909	3,776	3,625	3,697	3,771	3,847	3,924	4,002	4,082	4,164	4,247	4,332	4,419	4,507	4,597

Table 4-10. Estimated VOC Emissions (Tons): Proposed Option

	2008	2009	2010	2011	2012	2013	2014	2015	2016	2017	2018	2019	2020	2021	2022
Wood stoves	1,085	693	666	679	692	706	720	735	750	765	780	293	298	304	311
Single burn rate stoves	2,243	1,433	1,376	1,403	1,431	1,460	337	344	351	358	365	106	109	111	113
Pellet stoves	4	2	2	2	2	2	2	3	3	3	3	1	1	1	1
Furnace: indoor, cordwood	1,793	1,146	1,100	1,122	1,144	1,167	298	304	310	316	322	26	27	27	28
Hydronic heating systems	785	502	481	491	501	511	52	53	54	55	56	12	12	12	12
Total	5,909	3,776	3,625	3,697	3,771	3,847	1,410	1,438	1,467	1,496	1,526	438	447	456	465

Table 4-11. Estimated VOC Emissions (Tons): Alternative Option

	2008	2009	2010	2011	2012	2013	2014	2015	2016	2017	2018	2019	2020	2021	2022
Wood stoves	1,085	693	666	679	692	706	720	735	750	541	552	563	574	586	311
Single burn rate stoves	2,243	1,433	1,376	1,403	1,431	1,460	337	344	351	197	201	205	209	213	113
Pellet stoves	4	2	2	2	2	2	2	3	3	2	2	2	2	3	1
Furnace: indoor, cordwood	1,793	1,146	1,100	1,122	1,144	1,167	298	304	310	63	64	66	67	68	28
Hydronic heating systems	785	502	481	491	501	511	52	53	54	28	28	29	29	30	12
Total	**5,909**	**3,776**	**3,625**	**3,697**	**3,771**	**3,847**	**1,410**	**1,438**	**1,467**	**831**	**848**	**864**	**882**	**899**	**465**

From the tables above, we can show that the VOC emission reductions in 2022 are the same for each option at 4,132 tons. The average of the annual emission reductions between the year of rule promulgation, 2014, and the year that both options are fully implemented (2022) is 3,237 tons for the Proposed option and 3,241 tons for the Alternative option.

4.5 Methodology for Estimating CO Emissions from New Units

We used the same methodology described in Section 4.3 to develop estimates for CO emissions. Using the RWC database, we developed an estimate of CO emissions per appliance using baseline emission factors. Then, using the same NSPS phase-in assumptions and anticipated emission reductions (i.e., that CO reductions are comparable to $PM_{2.5}$ reductions), we developed emission factors to be used in analyzing the changes in emissions from applying the NSPS options. Table 4-12 presents the CO emission factors.

Table 4-12. NSPS CO Emission Factors

Emission Inventory Category	Baseline Emission Factor (lb/ton)	Baseline Emissions (tons)	Tons/Appl/Yr	Step 1 Emission Factor (lb/ton)	Step 1 Tons/Appl/Yr	Alternative Step 2 Emission Factor (lb/ton)	Alternative Step 2 Tons/Appl/Year	Alternative Step 3 Emission Factor (lb/ton)	Alternative Step 3 Tons/Appl./Year
Woodstove: fireplace inserts; EPA certified; non-catalytic	140.8	86,323	0.0662	140.8	0.0662	77.4	0.0364	40.269	0.0189
Woodstove: fireplace inserts; EPA certified; catalytic	104.4	21,725	0.0509	104.4	0.0509	104.4	0.0509	54.288	0.0264
Woodstove: freestanding, EPA certified, non-catalytic	140.8	108,418	0.1241	140.8	0.1241	77.4	0.0683	40.269	0.0355
Woodstove: freestanding, EPA certified, catalytic	104.4	40,486	0.1082	104.4	0.1082	104.4	0.1082	54.288	0.0563
Woodstove: pellet-fired, general	15.9	9,344	0.0110	15.9	0.0110	14.31	0.0099	7.441	0.0052
Single burn rate stoves (freestanding, non-EPA-certified)	230.8	249,785	0.2442	140.8	0.1489	77.4	0.0819	40.269	0.0426
Hydronic heater: outdoor	184	33,618	0.0922	18.4	0.0922	9.2	0.0461	3.680	0.0184
Furnace: indoor, cordwood	184	60,355	0.6878	46	0.1719	9.2	0.0344	3.680	0.0138

Using the same assumptions as we used for $PM_{2.5}$, we calculated CO emissions at baseline and for the two NSPS options we are considering. Table 4-13 shows the annual baseline emissions for CO for 2008 to 2022, and Tables 4-14 and 4-15 show the CO emissions under the Proposed and Alternative options.

Table 4-13. Estimated CO Emissions (Tons): Baseline

	2008	2009	2010	2011	2012	2013	2014	2015	2016	2017	2018	2019	2020	2021	2022
Wood stoves	10,918	6,976	6,697	6,831	6,968	7,107	7,249	7,394	7,542	7,693	7,847	8,004	8,164	8,327	8,494
Single burn rate stoves	9,766	6,241	5,991	6,111	6,233	6,358	6,485	6,614	6,747	6,882	7,019	7,160	7,303	7,449	7,598
Pellet stoves	1,438	919	882	900	918	936	955	974	994	1,014	1,034	1,055	1,076	1,097	1,119
Furnace: indoor, cordwood	28,198	18,019	17,298	17,644	17,997	18,357	18,724	19,098	19,480	19,870	20,267,	20,672	21,086	21,508	21,938
Hydronic heating systems	12,343	7,887	7,572	7,723	7,878	8,035	8,196	8,360	8,527	8,698	8,872	9,049	9,230	9,415	9,603
Total	**62,663**	**40,042**	**38,440**	**39,209**	**39,993**	**40,793**	**41,609**	**42,441**	**43,290**	**44,156**	**45,039**	**45,940**	**46,858**	**47,796**	**48,751**

Table 4-14. Estimated CO Emissions (Tons): Proposed Option

	2008	2009	2010	2011	2012	2013	2014	2015	2016	2017	2018	2019	2020	2021	2022
Wood stoves	10,918	6,976	6,697	6,831	6,968	7,107	7,249	7,394	7,542	7,693	7,847	2,743	2,797	2,853	2,910
Single burn rate stoves	9,766	6,241	5,991	6,111	6,233	6,358	3,956	4,035	4,116	4,198	4,282	1,249	1,274	1,300	1,326
Pellet stoves	1,438	919	882	900	918	936	955	974	994	1,014	1,034	494	503	513	524
Furnace: indoor, cordwood	28,198	18,019	17,298	17,644	17,997	18,357	4,681	4,775	4,870	4,967	5,067	413	422	430	439
Hydronic heating systems	12,343	7,887	7,572	7,723	7,878	8,035	820	836	853	870	887	181	185	188	192
Total	**62,663**	**40,042**	**38,440**	**39,209**	**39,993**	**40,793**	**17,661**	**18,014**	**18,375**	**18,742**	**19,117**	**5,080**	**5,181**	**5,285**	**5,391**

Table 4-15. Estimated CO Emissions (Tons): Alternative Option

	2008	2009	2010	2011	2012	2013	2014	2015	2016	2017	2018	2019	2020	2021	2022
Wood stoves	10,918	6,976	6,697	6,831	6,968	7,107	7,249	7,394	7,542	5,069	5,171	5,274	5,380	5,487	2,910
Single burn rate stoves	9,766	6,241	5,991	6,111	6,233	6,358	3,956	4,035	4,116	2,309	2,355	2,402	2,450	2,499	1,326
Pellet stoves	1,438	919	882	900	918	936	955	974	994	912	930	949	968	987	524
Furnace: indoor, cordwood	28,198	18,019	17,298	17,644	17,997	18,357	4,681	4,775	4,870	993	1,013	1,034	1,054	1,075	439
Hydronic heating systems	12,343	7,887	7,572	7,723	7,878	8,035	820	836	853	435	444	452	461	471	192
Total	**62,663**	**40,042**	**38,440**	**39,209**	**39,993**	**40,793**	**17,661**	**18,014**	**18,375**	**9,719**	**9,913**	**10,112**	**10,314**	**10,520**	**5,391**

From the tables above, we can show that the CO emission reductions in 2022 are the same for each option at 43,360 tons. The average of the annual CO emission reductions between the year of rule promulgation, 2014, and the year that both options are fully implemented (2022) is 32,559 tons for the Proposed option and 32,873 tons for the Alternative option.

SECTION 5
COST ANALYSIS, ENERGY IMPACTS, AND EXECUTIVE ORDER ANALYSES

In this section, we provide the estimates of total compliance costs and background behind their estimation. In addition, we provide a qualitative economic impact analysis of the proposed rule's impact on consumer and producer decisions, a qualitative discussion on unfunded mandates that may occur as a result of this final rule, and a partial analysis of the impacts of this proposal on employment. We used the direct annual compliance costs as an approximate measure of total social costs.

Given these constraints, several economic frameworks can be used to estimate the economic impacts and social costs of regulations; however, OAQPS has traditionally relied on partial equilibrium market models. Previous NSPS economic impact analyses for the residential wood stove market were prepared reflecting such a model standpoint. However, the current data do not provide sufficient details to develop a market model; the data that are available have little or no sector/firm detail and are reported at the national level. In addition, some sectors have unique market characteristics that make developing partial equilibrium models difficult. Therefore, we have prepared the economic impact analysis using a qualitative partial equilibrium framework.

5.1   Background for Compliance Costs

5.1.1   Estimated Research and Development (R&D) Costs

5.1.1.1 Residential Wood Heaters (except for masonry heaters)

EPA has received various estimates of the costs to bring a wood heater from concept to completion, from $300,000 for a single model to $1,360,000 for a 4-firebox model line. A recent Hearth and Home article estimated the total cost to bring a model from conception to market as $645,000 to $750,000 for steel stoves and over $1 million for cast-iron, enameled wood stoves. The authors indicated that costs would decrease for separate models in the same line by up to 25%. Based on this information, we estimate that a 4-model steel line would cost up to $328,125 per model to develop. These costs include marketing, design, developing first generation, second generation and prototype units; NSPS and safety testing, equipment tooling, etc. The manufacturer supplying these figures for the article estimated that the NSPS and safety testing component of these costs would constitute $40,000 per model. This manufacturer said that development time is 12 to 14 months for non-catalytic heaters and 10 to 12 months for catalytic heaters.[33]

Another manufacturer estimated costs of new product development, including design, prototype development, testing, tooling equipment and other manufacturing changes, marketing support, materials, training, and education to be in excess of $300,000 over an 8- to 12-month schedule for a relatively uncomplicated product. Costs will increase for products that have more sophisticated controls. [34] One other manufacturer estimated that their typical model development costs are around $200,000/model.[35]

Two manufacturers suggested a 14- to 18-month time frame is required to develop a new firebox, but added that it will take from 5 to 6 years of intensive engineering and R&D efforts to have a model line consisting of 4 boxes ready for manufacture. They agreed that knowledge of the process obtained during each firebox development will shorten (somewhat) the time necessary, but not enough to consider within a guiding framework. These manufacturers also provided estimated development costs for a 4-box model line, presented in Table 5-1 below.[36]

Table 5-1.   Example of Manufacturers' Estimates of Costs to Develop Model Line (4 Fireboxes)

Cost Component	Estimated Costs	Notes
Salaries	$850,000	Using 2-full time experimented employees to bring the products to market, salaries and benefits are estimated at $160,000 per year for at least 5.3 years. Tasks include design, prototyping, testing, production-line integration, and marketing.
Laboratory Equipment	$50,000	In order to accelerate R&D and avoid validating each result with independent testing labs (too costly for most manufacturers), new testing equipment will need to be purchased in order to sample flue gases, measure test load weight loss, record data automatically, and analyze flue gases composition.
Prototypes	$25,000	Numerous prototypes will be needed until the final product can be approved. For each firebox, estimate that 8 prototypes will be needed, at a cost of $700 each. Numerous samples of various components will also have to be purchased from vendors.
Test Fuel[a]	$45,000	Each test costs at least $50 in fuel (assuming cribs are used), including waste. An estimated 150 tests will have to be conducted for each firebox, for a total of $7,500, or $30,000 for a 4-firebox model line based on crib testing.

(continued)

[33] James E. Houck and Paul Tiegs. *There's a Freight Train Comin'*. Hearth and Home. December 2009.
[34] Comments from United States Stove Company, Small Entity Representative. July 13, 2010.
[35] Confidential Business Information.
[36] NSPS Review/Revision, and Impact on Our Companies: A Manufacturer's Position Statement. Prepared by Stove Builder International and United States Stove Company. June 2010.

Table 5-1. Example of Manufacturers' Estimates of Costs to Develop Model Line (4 Fireboxes) (continued)

Cost Component	Estimated Costs	Notes
Testing Services[a]	$150,000	Testing services for emissions, efficiency, and safety are estimated to last approximately 3 weeks for each firebox. At an average of $1,500 per day plus travel expenses, this amounts to approximately $25,000 for each firebox, or $100,000 for a 4-firebox model line based on crib testing.
Outside Consultants	$160,000	The average manufacturer will need outside help for design and testing. Testing equipment, knowledge of the test standard, and general guidance is normally offered by outside consultants (not necessarily certified EPA test labs). The average manufacturer will need approximately 300 hours of consulting services per year ($40,000) for 4 years.
Re-tooling	$120,000	For each firebox, new molds and jigs will need to be purchased or produced. Estimate that re-tooling charges will reach at least $30,000 per firebox, or $120,000 for a 4-firebox model line.
Marketing	$25,000	New pictures will need to be taken and all the current marketing material, including web sites and owner's manuals, will have to be updated.
Total	$1,360,000	Equal to $340,000/model

[a] Note: As described in our unit cost memo, the costs originally provided by industry for this table were presumed to be based on crib wood testing, not both crib wood and cord wood testing. Therefore we increased the industry-based "Test Fuel" cost by 50% (to the $45,000 shown above) as well as the industry-based "Testing Services" cost by 50% (to the $150,000 shown above) in order to estimate the additional cost to test with both crib wood and cord wood.

For this analysis, we used the costs provided in the Table 5-1 example, scaled to a single model and spread over a 6-year model development time frame. We prepared an annualized R&D cost estimate by separating cost elements into direct annual costs (salaries) vs. indirect annual costs (laboratory equipment, retooling and other capital costs). We estimated annual capital costs during the amortized R&D cost period as the fraction that the indirect costs (IAC) are of the Total Annual Cost, approximately 34% annually. Ongoing costs such as taxes, overhead, and other routine expenses would be incurred regardless of the NSPS standard, and are not included in this analysis. Table 5-2 shows the estimated annualized cost of $63,850 per model, assuming an amortization period of 6 years and an interest rate of 7%.

Table 5-2. Annual Cost Summary: Development of 4 Model Fireboxes[a,b]

Direct Annual Costs (DAC)		
Operator labor	$141,667	Annual salary cost from Table 1, spread over 6 years."
Outside Consultants	$26,667	Annual outside consultant cost from Table 1, spread over 6 years."
Total Direct Costs (DC)	$168,333	

(continued)

Table 5-2. Annual Cost Summary: Development of 4 Model Fireboxes[a,b] (continued)

Indirect Annual Costs (IAC)		
Laboratory Equipment[a]	$10,490	The laboratory equipment cost of $50,000 was amortized over 6 years at an interest rate of 7%.
Re-tooling[b]	$25,175	The re-tooling cost of $120,000 was amortized over 6 years at an interest rate of 7%.
Other Capital Costs[b]	$51,400	Other capital costs include costs for prototypes ($25,000), test fuel ($45,000), testing services ($150,000), and marketing ($25,000) and were amortized over 6 years at an interest rate of 7%.
Total Indirect Costs (IAC)	$87,065	
Total Annual Cost	$255,399	Annual cost for development of 4 model fireboxes.
Total Annual Cost	$63,850	Average annual cost per firebox model.

[a] An amortization period of 6 years for laboratory equipment, retooling and other capital costs was chosen based on industry's estimate that approximately 5 to 6 years of R&D are required to bring a product to market.

[b] As described in the unit cost memo, to estimate the additional cost to test with both cord wood and crib wood, the test fuel industry estimate of $30,000 based on crib only was increased by $15,000 and the testing services industry estimate of $100,000 based on crib only (which covered not only emissions testing but also efficiency and safety testing) was increased by $50,000.

5.1.2 Masonry Heaters

Masonry heaters manufacturing cost impacts vary by the type of producer and the type of certification method. According to one manufacturer,[37] the masonry heater industry in the U.S. is dominated by the Finnish firm Tulikivi, which manufactures and imports about half of the U.S. masonry heater units installed yearly through its network of installing distributors. The same manufacturer said that the second largest producer is a Canadian firm, Temp-Cast. The remainder of the industry is made up of "dozens" of small producers, with probably fewer than 100 (or at least fewer than 200) generating any masonry revenue at all. Some commercial operations sell core units and/or design kits based on their own design, and other sell units they license from other U.S. or foreign companies. Finally, some units are custom built. Based on this information, we assumed that 50% of masonry heaters sold per year in the U.S. are Tulikivi models and 35% are sold by other manufacturers. The remaining 15% of units are sold by independent contractors.

There are three major cost components to consider in evaluating the potential cost impacts of the proposed NSPS: research and development (R&D), certification testing, and licensing fees for use of a computer software package approved for use in certifying a model

---

[37] Comments: Residential Solid Fuel Burning Appliance SBREFA Process. Product Category: Masonry Heaters. July 13, 2010. Timothy Seaton, Timely Construction, Inc. p. 5.

design. According to information provided by one manufacturer, capital R&D costs for a masonry heater may be estimated at $250,000.[38] In the absence of more specific data regarding R&D costs for masonry heaters, we assumed R&D costs were the same as for other wood heater appliances—that is, $63,850 annually for a 6-year R&D amortization period. For facilities conducting R&D, these costs include the costs for certification testing. We estimate that the cost of testing a heater design in an EPA accredited lab to be approximately $10,000.[39]

This cost analysis also makes use of a unique software package based on a European masonry heater design standard. This standard has been verified in the laboratory and under field conditions to produce masonry heaters that would meet the proposed NSPS emission limits. The software produces for printout a certification for a given design application and the design definition documents as well as operating instructions customized to the given design, so that the software verification and certification record is created for and attached to the design. The resulting documents can be submitted as part of the certification application. The cost of this software to the user is approximately 1,000 Euros (approximately $1,500) for the package with a 300 Euro (approximately $450) annual fee that commences in the second year following purchase.[40]

5.1.3 General Approach and Assumptions for Costs to Manufacturers

Manufacturers have told us that it takes several years to develop new models, and this is documented in the manufacturer's cost memo.[41] We have spread the annualized R&D costs (shown in Table 5-2) over 6 years to best represent the time and funds needed to develop the complying models. For the purposes of our cost estimate, we have assumed that when the NSPS revisions are proposed, all manufacturers will begin serious efforts to develop complying models, although for some manufacturers we also know that they have been involved in intensive R&D efforts in anticipation of the proposed rule.

We estimated both the average annual cost to manufacturers of each appliance type and then extended those costs to nationwide total annual costs. The basic components to each manufacturer's estimated annual cost are:

---

[38] Comments: Residential Solid Fuel Burning Appliance SBREFA Process. Product Category: Masonry Heaters. July 13, 2010. Timothy Seaton, Timely Construction, Inc. p. 14.

[39] Letter to Lucinda Power, EPA, from Brian Klipfel, Fire Works Masonry. September 10, 2010.

[40] E-mail from Timothy Seaton, Timely Construction Company, to Gil Wood, USEPA. April 21, 2011.

[41] [41] Memo to Gil Wood, USEPA, from Jill Mozier, EC/R, Inc. Estimated Wood Heater Manufacturer Cost Impacts. February 22, 2013.

- Annualized R&D cost;

- Ongoing annual Certification cost; and

- Ongoing annual Reporting and Record Keeping cost.

The Annualized R&D costs (shown in Table 5-2, and based on the Table 5-1 costs) are by far the largest cost component and we have applied these costs to most models in our cost analysis—especially to models in previously unregulated appliance categories—in order to present a reasonable estimate of the costs given uncertainty over the precision of available estimate of the R&D cycle lifespan. For example, as noted above, instead of estimating the number of hydronic heater models that already meet a specific limit and will therefore merely need to certify their emissions rather than undergo R&D, we instead assumed that 100% of hydronic heater models will undergo R&D beginning in 2013. We made similar reasonable assumptions for single burn rate stoves and forced air furnaces.

Under the Proposal scenario, one round of R&D is assumed—beginning in 2013 and ending in 2018—in order to meet the proposed Step 2 limit. Under the Alternative option, two rounds of R&D are assumed for all appliances except masonry heaters (for which there is only one standard with no additional phased-in standards to meet). Under the Alternative option, the first R&D round begins in 2013 and the second round begins in 2017 (which causes overlapping R&D costs in years 2017 and 2018 in this analysis)—in order to meet the interim Alternative Step 2 limit in 2017 and the Alternative Step 3 limit in 2022. We also reasonably assumed that of the models undergoing the first round of R&D costs, 80% of these models undergo and assume the second round of R&D costs in the Alternative scenario (i.e., we reasonably assumed that only 20% of models achieve the strictest limit in the first round of R&D).

Furthermore, for appliances like single burn rate stoves and forced air furnaces, which were previously unregulated (and also were not pushed technologically by a voluntary program, as hydronic heaters were), we have reasonably doubled R&D costs during years 2013 and 2014. This doubling of R&D cost estimates is to represent an intensification of the R&D efforts to meet the Step 1 limit and begin development of models to meet the stricter stepped limits—R&D efforts which industry has indicated are already ongoing.[42]

---

[42] 2/8/13 telephone discussion between Gil Wood, USEPA, and a manufacturer of forced air furnaces and single burn rate stoves.

Note that all manufacturers, except for wood stoves that are subject to the current 1988 NSPS, will face ongoing certification costs above baseline conditions. However, in the 2013 to 2018 time frame under the Proposal option and in the 2013 to 2022 time frame under the Alternative option, we have incorporated these costs as part of the overall R&D expenditures. After 2018 under the Proposal option and after 2022 under the Alternative option, the ongoing certification costs will be the only NSPS related costs faced by manufacturers besides ongoing reporting and recordkeeping costs.

Regarding certification costs, we have assumed a cost of $10,000 per model for pellet stoves, single burn rate stoves and masonry heaters; and we have assumed a cost of $20,000 per model for hydronic heaters and forced air furnaces.[43] We have spread these costs out over the 5 year certification life, assuming annual certification costs for one-fifth of the models.

For example, pellet stoves will incur certification costs in advance of complying with more stringent limits. As explained in the manufacturer's cost memo,[44] approximately 30% of existing pellet stove models are expected to comply with the proposed Step 2 and Alternative Step 3 standard. However, in order to be sold, these stove models would now be required to demonstrate compliance with an emissions limit, incurring an upfront cost of $10,000 per model to become certified. The same cost memo also discusses our assumption that one fifth of the pellet stove models will certify in any given year.

We based reporting and recordkeeping (R&R) costs on the annual average costs derived from development of the Information Collection Request (ICR) supporting statements.[45] These are reasonable annual estimates of the ongoing R&R burden to manufacturers associated with the Proposal and Alternative scenarios. (We do not expect the R&R burden to differ substantially between the two scenarios.)

The certification and reporting and recordkeeping costs were estimated to be incurred by manufacturers for the full 20-year model design lifespan.[46] Under the Proposal, we estimated costs from 2013 through 2038—that is, 20 years after the 2019 compliance year marking the beginning of the model lifespan designed to meet the Proposal Step 2 limit. Under the

---

[43] Conversation with Dennis Brazier, Central Boiler. August 9, 2010.
[44] Memo to Gil Wood, USEPA, from Jill Mozier, EC/R, Inc. Estimated Wood Heater Manufacturer Cost Impacts. February 22, 2013.
[45] ICR Supporting Statements for the Proposed NSPS Subparts have not been finalized as of the date of this memo.
[46] Memo to Gil Wood, USEPA, from Jill Mozier, EC/R, Inc. Unit Cost Estimates of Residential Wood Heating Appliances. February 21, 2013.

Alternative, we estimated costs from 2013 through 2041—that is, 20 years after the 2022 compliance year marking the beginning of the model lifespan designed to meet the Alternative Step 3 limit. We provide these costs in the manufacturer's cost memo.

5.1.4 General Approach and Assumptions for Costs to Masonry Heater Manufacturers

As noted above, we addressed masonry heaters in a way which segmented the costs in keeping with the masonry heater market. There are three scenarios for potential cost impacts for large masonry heater manufacturers. In the case of Tulikivi and some U.S. firms, e.g., Timely Construction, these companies have already invested in R&D in order to gain access to U.S. markets which restrict sales (e.g., Colorado) of uncertified units. These companies will face testing costs only, with an assumed total of nine tests conducted prior to the proposed compliance date (i.e., to certify a total of nine model lines). For purposes of our cost analysis, we assumed as shown in the unit cost memo that two additional companies will conduct R&D to develop two new models each to meet the proposed NSPS.[47] Finally, we have been told that Tulikivi will use the software certification approach to certify up to eight additional models. We also project as presented in the unit cost memo that the remaining 15% of custom built units will use the software certification approach to certify compliance with the proposed NSPS starting in 2013 (estimated date of the proposed standards) and that they will continue to renew their license in the following years.

As explained in the unit cost memo, we used data in the Frost & Sullivan Market (F&S) report[48] on 2008 masonry heater shipments by product category and F&S revenue forecasts which incorporated the weak economy in years 2009 and 2010, to calculate the reduced number of shipments in years 2009 and 2010. For years 2011 through 2038 (for the Proposal option) and 2011 through 2041 (for the Alternative option) estimated shipments are based on a forecasted revenue growth rate of 2.0%, in keeping with the average annual growth in real GDP predicted by the Conference Board.[49] For masonry heaters, our estimate of the number of custom built models is based on the average number of models sold per year in the 15% model category (i.e., 85 per year). We assumed each custom manufacturer would sell 2 models per year, for a total of 42 manufacturers participating in the software certification option.

---

[47] U.S. EPA. Memorandum. Unit Cost Estimates of Wood Heating Appliances. February 21, 2013. Prepared by EC/R, Inc.

[48] Market Research and Report on North American Residential Wood Heaters, Fireplaces, and Hearth Heating Products Market. Prepared by Frost & Sullivan. April 26, 2010. P. 31-32.

[49] 2013 Global Outlook projections prepared by the Conference Board in November 2012; http://www.conference-board.org/data/globaloutlook.cfm.

Under both the Proposal and Alternative options, most sales-weighted masonry heater units face a 2014 Step 1 compliance date with no other phased-in limits. However, under both the Proposal and Alternative options, companies that sell fewer than 15 units per year have until 2019 to come into compliance. We have assumed that the large manufacturers will comply by 2014 for the units that only require testing and/or software certification, with those expenditures incurred annually starting in 2013. We also assumed that the 15% of custom built units will comply by 2019, but will begin certifying their units using the software certification approach as early as 2013, as noted above, as a selling point for their services. For those companies that start R&D when the NSPS is proposed in 2013, we have assumed that they will spread these costs over the 6-year period from 2013 through 2018 for the four models affected, under both the Proposal and Alternative options.

5.1.5   General Approach and Assumptions for All Appliances

Below is a list of approach and assumptions for estimating costs for each category of appliances affected by this proposal, as taken from the manufacturer cost memo:[50]

1. Nationwide Annual Cost assumes R&D investment is amortized over 6 years (2013 through 2018). Ongoing certification costs are incurred through 2038 (based on a model brought to market in 2019 with a lifespan of 20 years), except for woodstoves which already incur certification costs under the existing NSPS.

2. Since certification is required every 5 years (except for the software certification option for masonry heaters), it is assumed that certification costs will be spread out so that 1/5 of the models certify each year.

3. This analysis considers additional costs resulting from the proposed NSPS. For wood stoves, the analysis assumes that 5% meet Step 2 already so that 95% of the models will undergo re-design to meet the Step 2 level. The costs modeled for years 2020 through 2038 exclude the ongoing certification costs and ongoing reporting and recordkeeping costs incurred by wood stove manufacturers who already had to certify and report under the existing NSPS.

4. For pellet stoves, the analysis assumes that 30% meet Step 2 already so that 70% of models undergo R&D re-design to meet Step 2. The R&D budget includes certification costs. The analysis also assumes that the 30% of the pellet stove models which already meet Step 2 will certify in an ongoing basis starting in 2013. The analysis reflects the certification costs beginning in 2013 for the 30% of models meeting Step 2, and beginning in 2019 for the remaining 70% of models which underwent R&D re-design.

---

[50] U.S. EPA. Memorandum. Estimated Residential Wood Heater Manufacturer Cost Impacts. February 22, 2013. Prepared by EC/R, Inc.

5. Based on conversations with industry in February 2013, single burn rate stoves and forced air furnaces have been undergoing R&D prior to 2013 to develop cleaner models. Because these devices were previously unregulated and may need to transfer technology from adjustable burn rate stoves and hydronic heaters respectively, this analysis assumes that these efforts will intensify in 2013 and 2014. Therefore estimated R&D costs are doubled in 2013 and 2014 in order to meet the 2014 Step 1 standard while also beginning R&D to develop models to meet the more stringent 2019 Step 2 standard.

6. For single burn rate stoves, forced air furnaces, and hydronic heating systems, the analysis assumes that only a small percentage meet Step 2 so that approximately 100% of the models undergo R&D re-design to meet Step 2. The R&D budget includes certification costs. Ongoing certification costs for the re-designed models are reflected in this analysis beginning in 2019.

7. For masonry heaters, the cost analysis assumes one round of R&D to meet 0.32 lb/mmBTU standard (no additional stepped standards, although large manufacturers will be required to meet the limit in 2014, while small volume manufacturers will be given a 5 year extension until 2019 to meet the limit). For masonry heater manufacturers using software certification, the analysis assumes the purchased software will be used for certifying all models developed by that manufacturer.

8. Reporting and recordkeeping costs (R&R) [for all appliances but masonry heaters] are based on the annual average costs derived from the ICR and are estimates of the ongoing R&R burden to manufacturers associated with the proposed NSPS. The annual average nationwide R&R burden estimated to manufacturers for Subpart AAA is $440,443, and for Subpart QQQQ is $119,249. These R&R costs do not include the R&R burden to laboratories; the annual average nationwide R&R burden incurred by laboratories subject to requirements under Subpart AAA is estimated to be $75,745, and incurred by laboratories subject to requirements under Subpart QQQQ is estimated to be $50,496.

9. For Masonry Heaters, Reporting and recordkeeping costs (R&R) are based on the annual average costs derived from the ICR and are estimates of the ongoing R&R burden to manufacturers associated with the proposed NSPS. The annual average nationwide R&R burden estimated to manufacturers for Subpart RRRR is $98,788 for small/custom masonry heater manufacturers and $25,929 for large masonry heater manufacturers. These R&R costs do not include the R&R burden to laboratories; the annual average nationwide R&R burden incurred by laboratories subject to requirements under Subpart RRRR is estimated to be $37,872.

For the Alternative option for all appliances, here are the assumptions for cost estimations where they differ from those in the Proposed option:

1. Nationwide Annual Cost assumes R&D investment is amortized over 6 years (round one from 2013 through 2018 and round two from 2017 through 2022). Ongoing certification costs are incurred through 2041 (based on a model brought to market in

2022 with a lifespan of 20 years), except for woodstoves which already incur certification costs under the existing NSPS.

2. (Same as above)

3. This analysis considers additional costs resulting from the proposed NSPS. For wood stoves, the analysis assumes that 5% meet Step 3 already so that 95% of the models will undergo re-design in round one, and 80% of those 95% will require another round of R&D to meet the Step 3 level. The costs exclude the ongoing certification costs and ongoing reporting and recordkeeping costs incurred by wood stove manufacturers who already had to certify and report under the existing NSPS.

4. For pellet stoves, the analysis assumes that 30% meet Step 3 already so that 70% of models undergo re-design in round one, and 80% of those 70% require another round of R&D to meet Step 3. The R&D budget includes certification costs. The analysis also assumes that the 30% of the pellet stove models which already meet Step 3 will certify in an ongoing basis starting in 2013.

5. Based on conversations with industry in February 2013, single burn rate stoves and forced air furnaces have been undergoing R&D prior to 2013 to develop cleaner models. Because these devices were previously unregulated and may need to transfer technology from adjustable burn rate stoves and hydronic heaters respectively, this analysis assumes that these efforts will intensify in 2013 and 2014. Therefore estimated R&D costs are doubled in 2013 and 2014 in order to meet the 2014 Step 1 standard while also beginning R&D to develop models to meet the more stringent 2017 Step 2 and 2022 Step 3 standards.

6. For single burn rate stoves, forced air furnaces, and hydronic heating systems, the analysis assumes that only a small percentage meet Step 3 so that approximately 100% of the models undergo re-design in round one, and 80% require another round of R&D to meet Step 3. The R&D budget includes certification costs.

5.1.6   Labor Requirements for Monitoring, Recordkeeping and Reporting

The focus of this part of the analysis is on labor requirements related to the compliance actions of the affected entities within the affected sector. This analysis estimates the labor requirements associated with new reporting and recordkeeping requirements.

The labor changes may either be required as part of an initial effort to comply with the new regulation or required as a continuous or annual effort to maintain compliance. We estimate up-front and continual, annual labor requirements by estimating hours of labor required for the monitoring, recordkeeping, and reporting efforts to maintain compliance.

The results of this analysis are presented in Table 5-3 for the Proposed NSPS option. The table breaks down the certification, quality assurance, reporting and recordkeeping burden and labor estimates by appliance type for each of the proposed subparts and for the test labs to obtain and maintain testing accreditation. These estimates are presented in terms of the estimated hours required. These estimates are consistent with estimates EPA submitted as part of the Information Collection Requests (ICRs) that are in the Supporting Statements for the proposed rules.

We note that certification testing (once every 5 years unless a waiver is granted) costs of approximately $10,000 ($20,000 for hydronic heaters and forced air furnaces) per model line in 2022 result in labor hours spent at the test lab, which are not included in this labor rate analysis. In addition, each model that is developed (i.e., number of affected units) will face an annual estimated cost of $160,000 for the salaries of two full-time experimental employees for 5 years. This estimate should be regarded as a partial labor estimate, with other possible labor associated with new model development (such as re-tooling) left as unquantified and described qualitatively in the manufacturer cost and unit cost impact memos.

Ongoing labor requirements are estimated at about 9,900 hours for the Proposed option. The labor estimate for the Alternative option will not be substantially different to that for the proposed option since there are no additional labor requirements for administrative matters under the Alternative option compared to the Proposed option.[51] These ongoing labor requirements can be viewed as average sustained labor requirements required for affected entities to continuously comply with the new regulations from 2019 and beyond.

---

[51] Memo to Gil Wood, USEPA, from EC/R, Inc. Residential Wood Heating Cost Effectiveness Analysis. February 26, 2013.

Table 5-3. Estimates of Labor Requirements for Certification, Quality Assurance, Reporting, Recordkeeping, and Accreditation Requirements for the Proposed NSPS Option[a]

Source/Emissions Point	Projected No. of Affected Units	Per-Unit One-Time Labor Estimate (Hours)	Total One-Time Labor Estimate (Hours)	Total Annual Labor Estimate (Hours)
Adjustable burn rate stoves not needing certification testing	125	0	0	1,784
Single burn rate stoves	20	0	0	283
Pellet stoves	125	0	0	1,779
Test Labs	6	0	0	597
Removable Label				1,017
Hydronic heaters—model development	90	0	0	1,290
Forced air furnaces—model development	38	0	0	539
Test Labs	4	0	0	398
Removable Label				332
Existing models at large manufacturers certified via testing	13	0	0	187
Existing models at large manufacturers certified via computer simulation	8	0	0	113
Existing models at small manufacturers certified via computer simulation	85	0	0	1,275
Test Labs	3	0	0	298.5
Removable Label				4.46
Total for Industry	**517**	**0**	**0**	**9,893**

[a] The labor requirements for the Alternative option are not substantially different than those for the Proposed option.

Note: Totals may not sum due to independent rounding. The Agency assumes in its cost analysis of monitoring, recordkeeping, reporting, and testing requirements in the Information Collection Request (ICR) that only half of the currently available models for all appliance types would be certified and sold.

5.2   Compliance Costs of the Proposed Rule

EPA's engineering cost analysis estimates that the total annualized cost of the proposed rule option to manufacturers for the Proposed option is $15.7 million, calculated as an average of the annualized costs incurred from 2014 to 2022. For the alternative option, the total annualized cost is $28.3 million (all costs are reported in 2010 dollars) (EC/R, December 2011) Annualized costs are estimated at a 7% interest rate.[52] We calculate the costs in this way in order to provide an average of annualized costs for these options from the time of rule promulgation in 2014 to the time of full implementation of both options, which occurs in 2022. Having an average annualized calculation for the costs allows for a reasonable measure of the costs to be incurred by manufacturers given the changes in costs year by year between 2014 and 2022 as shown in the manufacturer's cost memo for this proposal. The total annualized costs for each year and for each option are in Table 5-4.

With total annualized costs estimated at a 3% interest rate, the total annualized cost of the proposal option is $14.8 million, and $26.9 million for the alternative option (in 2010 dollars). More detailed information on the costs at a 3% interest rate can be found in the cost memorandum for this proposal.

Under the proposed NSPS option, the costs in the 2014-2022 time frame fall most heavily on manufacturers of hydronic heating systems (29%), followed by wood stoves (27%), then by pellet stoves (22%). The remaining 22% of the costs are distributed to forced-air furnace manufacturers (14%), manufacturers of single burn rate stoves (6%), and masonry heaters (2%).

---

[52] EC/R, Inc. to U.S. EPA, Draft Memorandum. Estimated Residential Wood Heater Manufacturer Cost Impacts. February 22, 2013.

Table 5-4. **Summary of Average Annualized Nationwide Costs for 2014–2022 Time Frame Under the Proposal and Alternative Options**

Appliance Type	Proposed Option	Alternative Option
Wood Heaters	$4,212,303	$8,090,026
Single Burn Rate Heaters	$901,732	$1,540,600
Pellet Stoves	$3,460,489	$6,255,536
Forced-Air Furnaces	$2,252,284	$3,813,898
Hydronic Heating Systems	$4,554,152	$8,302,026
Masonry Heaters	$307,511	$307,511
Total Average Annual Cost for 2014–2022 Time Frame	$15,688,471	$28,309,597

For the alternative option, the annualized costs fall most heavily of manufacturers of hydronic heaters (29%) and wood stoves (29%), followed by pellet stoves (22%). The remaining 20% of the costs are distributed to manufacturers of forced air heaters (13%), single burn rate stoves (5%), and masonry heaters (2%).

The revised rule, as proposed would affect an estimated 2.7 million new residential wood heating devices between 2014 and 2022 assuming an average of ~296,000 new shipments annually as presented in the emissions memo for the proposal.[53] As shown previously in Table 4-4 in Section 4, annual shipments are forecasted to increase for all product types over the same time period.

To assess the size of the compliance costs relative to the value of shipments to end-use consumers, we compared industry-level compliance costs relative to projected sales for 2018 since this is year between 2014 and 2022 and is a representative year suitable for this analysis. In this case, cost-to-receipts ratios approximate the maximum price increase needed for a producer to fully recover the annual compliance costs associated with a regulation. These industry-level cost-to-receipts ratios can be interpreted as an average impact on potentially affected firms in these industries all other things equal, and where ratios less than 1% suggest the rule will not have a significant impact using EPA's SBREFA guidance as a basis. Results for affected industries for the 2014–2022 time frame can be found in Tables 5-4a and 5-4b.

---

[53] U.S. EPA. Memorandum. Estimated Emissions from Wood Heaters. February 15, 2013. Prepared by EC/R, Inc.

Under the NSPS proposal option, none of the six affected product types would have an annualized cost-to-receipts ratio of less than 1%. In the 2014-2022 time frame for this option, cost-to-receipts ratios range from 2.3% for pellet stoves to 6.4% for single burn rate stoves as shown in Table 5-5a. For the alternative option, cost-to-receipts ratios range from 4.0% for forced air furnaces to 10.7% for single burn rate stoves as shown in Table 5-5b.

Table 5-5a. Industry Level-Annualized Compliance Costs (2010 dollars) as a Fraction of Total Industry Revenue by Product Type in the 2014–2022 Time Frame— Proposal Option

Product Type	Total Annualized Costs ($ millions)	Product Sales in 2018 ($ millions)[a]	Cost-to-Receipts Ratio
Wood stoves	$4.2	$98.1	4.3%
Single burn rate stoves	$0.9	$14.0	6.4%
Pellet stoves	$3.5	$152.8	2.3%
Forced-air furnaces	$2.3	$96.6	2.4%
Masonry heaters	$0.3	$6.3	4.8%
Hydronic heating systems	$4.5	$134.4	3.3%

[a] Sales based on projected product shipments and average unit costs estimates. We use annual sales in 2018 to approximate annual sales for years from 2014 to 2022. Total annualized costs in this table are estimated at a 7% interest rate.

Sources: Masonry Heater Compliance Costs from *Masonry Heater NSPS Annual Cost 12 8 11.xls*. Received from EPA on December 16, 2011.

Unit Costs and Shipment Projections from *Unit Cost Memo*. Received by EPA in February, 2013.

Industry Compliance Costs from *Wood Stove NSPS Annual Costs*. Received by EPA February, 2013.

Table 5-5b. Industry Level-Annualized Compliance Costs (2010 dollars) as a Fraction of Total Industry Revenue by Product Type in 2014–2022 Time Frame— Alternative Option

Product Type	Total Annualized Costs ($ millions)	Product Sales in 2018 ($ millions)[a]	Cost-to-Receipts Ratio
Wood stoves	$8.1	$98.1	8.3%
Single burn rate stoves	$1.5	$14.0	10.7%
Pellet stoves	$6.2	$152.8	4.1%
Forced-air furnaces	$3.8	$96.6	4.0%
Masonry heaters	$0.3	$6.3	4.8%
Hydronic heating systems	$8.3	$134.4	6.1%

[a] Sales based on projected product shipments and average unit costs estimates. We use annual sales in 2018 to approximate annual sales for years from 2014 to 2022. Total annualized costs are estimated at a 7% interest rate.

Sources: Masonry Heater Compliance Costs from *Masonry Heater NSPS Annual Cost 12 8 11.xls*. Received from EPA on December 16, 2011.

Unit Costs and Shipment Projections from *Unit Cost Memo*. Received by EPA in February, 2013.

Industry Compliance Costs from *Wood Stove NSPS Annual Costs*. Received by EPA February, 2013.

5.3 How Might People and Firms Respond? A Qualitative Partial Equilibrium Analysis

Markets are composed of people as consumers and producers acting as economic agents to maximize utility or profits, respectively. One way economists illustrate behavioral responses to pollution control costs is by using market supply and demand diagrams. The market supply curve describes how much of a good or service firms are willing and able to sell to people at a particular price; we often draw this curve as upward sloping because some production resources are fixed. As a result, the cost of producing an additional unit typically rises as more units are made. The market demand curve describes how much of a good or service consumers are willing and able to buy at some price. Holding other factors constant, the quantity demanded is assumed to fall when prices rise. In a perfectly competitive market, equilibrium price (P_0) and quantity (Q_0) are determined by the intersection of the supply and demand curves (see Figure 5-2).

5.3.1 Changes in Market Prices and Quantities

To qualitatively assess how the regulation may influence the equilibrium price and quantity in the affected markets, we assumed the market supply function shifts up by the additional cost of producing the good or service; the unit cost increase is typically calculated by dividing the annual compliance cost estimate by the baseline quantity (Q_0) (see Figure 5-2). As shown, this model makes two predictions: the price of the affected goods and services are likely to rise and the consumption/production levels are likely to fall.

The size of these changes depends on two factors: the size of the unit cost increase (supply shift) and differences in how each side of the market (supply and demand) responds to changes in price. Economists measure responses using the concept of price elasticity, which represents the percentage change in quantity divided by the percentage change in price. This dependence has been expressed in the following formula:[54]

$$\textit{Share of per-unit cost} = \frac{\textit{Price Elasticity of Supply}}{(\textit{Price Elasticity of Supply - Price Elasticity of Demand})}$$

As a general rule, a higher share of the per-unit cost increases will be passed on to consumers in markets where

- goods and services are necessities and people do not have good substitutes that they can switch to easily (demand is inelastic) and

- suppliers have excess capacity and can easily adjust production levels at minimal costs, or the time period of analysis is long enough that suppliers can change their fixed resources; supply is more elastic over longer periods.

---

[54] For examples of similar mathematical models in the public finance literature, see Nicholson (1998), pages 444–447, or Fullerton and Metcalf (2002).

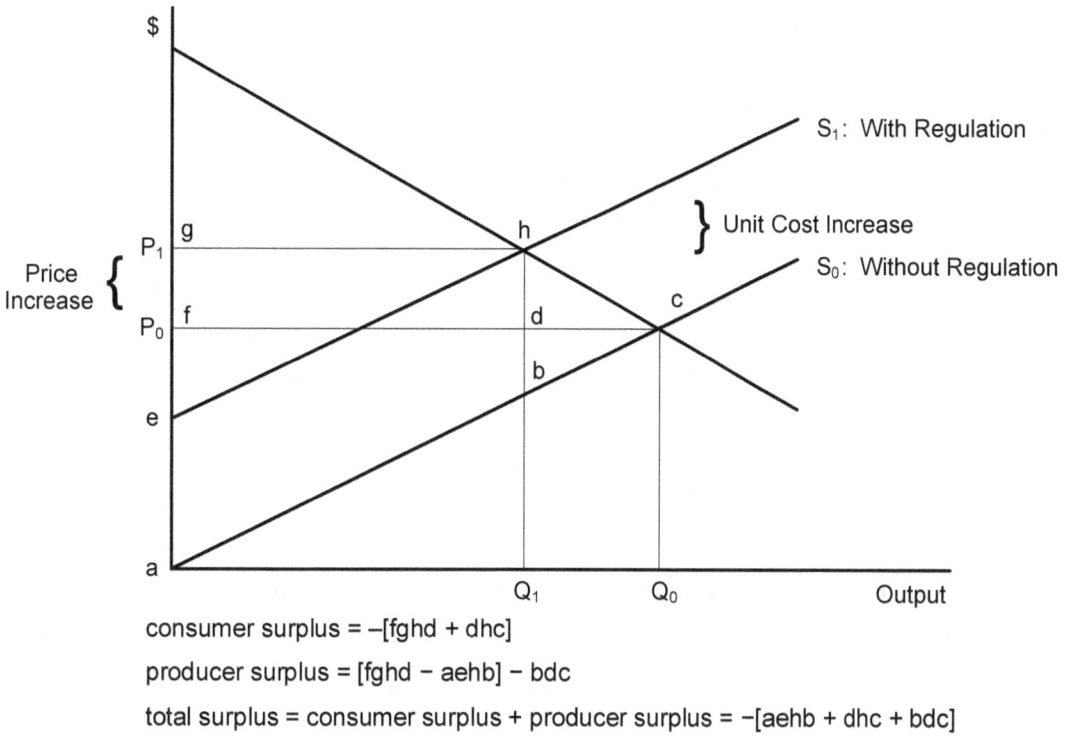

consumer surplus = –[fghd + dhc]

producer surplus = [fghd – aehb] – bdc

total surplus = consumer surplus + producer surplus = –[aehb + dhc + bdc]

Figure 5-2.   Market Demand and Supply Model: With and Without Regulation

Short-run demand elasticities for energy goods (electricity and natural gas), agricultural products, and construction are often inelastic. Specific estimates of short-run demand elasticities for these products can be obtained from existing literature. For the short-run demand of energy products, the National Energy Modeling System (NEMS) buildings module uses values between 0.1 and 0.3; a 1% increase in price leads to a 0.1 to 0.3% decrease in energy demand (Wade, 2003). For the short-run demand of agriculture and construction, EPA has estimated elasticities to be 0.2 for agriculture and approximately 1 for construction (EPA, 2004). As a result, a 1% increase in the prices of agriculture products would lead to a 0.2% decrease in demand for those products, while a 1% increase in construction prices would lead to approximately a 1% decrease in demand for construction. Given these demand elasticity scenarios (shaded in gray), approximately a 1% increase in unit costs would result in a price increase of 0.1 to 1% (Table 5-6). As a result, 10 to 100% of the unit cost increase could be passed on to consumers in the form of higher goods/services prices. This price increase would correspond to a 0.1 to 0.8% decline in consumption in these markets (Table 5-7).

Table 5-6.   Hypothetical Price Increases for a 1% Increase in Unit Costs

Market Demand Elasticity	Market Supply Elasticity						
	0.1	0.3	0.5	0.7	1	1.5	3
−0.1	**0.5%**	0.8%	0.8%	0.9%	0.9%	0.9%	1.0%
−0.3	0.3%	**0.5%**	0.6%	0.7%	0.8%	0.8%	0.9%
−0.5	0.2%	0.4%	**0.5%**	0.6%	0.7%	0.8%	0.9%
−0.7	0.1%	0.3%	0.4%	**0.5%**	0.6%	0.7%	0.8%
−1.0	0.1%	0.2%	0.3%	0.4%	**0.5%**	0.6%	0.8%
−1.5	0.1%	0.2%	0.3%	0.3%	0.4%	**0.5%**	0.7%
−3.0	0.0%	0.1%	0.1%	0.2%	0.3%	0.3%	**0.5%**

Table 5-7.   Hypothetical Consumption Decreases for a 1% Increase in Unit Costs

Market Demand Elasticity	Market Supply Elasticity						
	0.1	0.3	0.5	0.7	1	1.5	3
−0.1	**−0.1%**	−0.1%	−0.1%	−0.1%	−0.1%	−0.1%	−0.1%
−0.3	−0.1%	**−0.2%**	−0.2%	−0.2%	−0.2%	−0.3%	−0.3%
−0.5	−0.1%	−0.2%	**−0.3%**	−0.3%	−0.3%	−0.4%	−0.4%
−0.7	−0.1%	−0.2%	−0.3%	**−0.4%**	−0.4%	−0.5%	−0.6%
−1.0	−0.1%	−0.2%	−0.3%	−0.4%	**−0.5%**	−0.6%	−0.8%
−1.5	−0.1%	−0.3%	−0.4%	−0.5%	−0.6%	**−0.8%**	−1.0%
−3.0	−0.1%	−0.3%	−0.4%	−0.6%	−0.8%	−1.0%	**−1.5%**

5.3.2   Partial Equilibrium Measures of Social Cost: Changes in Consumer and Producer Surplus

In partial equilibrium analysis, the social costs are estimated by measuring the changes in consumer and producer surplus, and these values can be determined using the market supply and demand model (as shown in Figure 5-2). Assuming linear market supply and demand curves as shown in Figure 5-2, the change in consumer surplus (CS) is measured as follows:

$$\Delta CS = - [\Delta Q_1 \times \Delta p] + [0.5 \times \Delta Q \times \Delta p]. \qquad (5.1)$$

where a coefficient of 0.5 is multiplied to the change in Q and P resulting from the shock to the markets based on the assumption of linear demand and supply curves in the diagram above and applying principles of basic geometry. Higher market prices and lower quantities lead to consumer welfare losses. Similarly, the change in producer surplus (PS) is measured as follows:

$$\Delta PS = [\Delta Q_l \times \Delta p] - [\Delta Q_l \times t] - [0.5 \times \Delta Q \times (\Delta p - t)]. \tag{5.2}$$

Higher unit costs and lower production levels reduce producer surplus because the net price change ($\Delta p - t$) is negative. However, these losses are mitigated because market prices tend to rise.

5.4   Social Cost Estimate

As shown in Tables 5-5a and 5-5b, the social cost as approximated by the annual compliance costs as a percent of sales represent a fraction of the affected product value that is greater than 1% for each of the product categories; this suggests that the shift of the supply curve may be relatively large for some product types and result in larger changes in market prices and consumption. EPA believes the national annualized compliance cost estimates provide a reasonable approximation of the social cost of this proposed rule. EPA believes this approximation is better for industries whose markets are well characterized as perfectly competitive. However, given the data limitations noted earlier, EPA believes the accounting for annual compliance costs is a reasonable approximation to inform policy discussion in this rulemaking. We were not able to prepare a full economic analysis of the impacts of this proposal on supply and demand, or the effects of such impacts on emissions (e.g. feedback effect on emissions). Most of the affected industries can be characterized as having a high degree of competitive market behavior. To shed more light on the level of market behavior, EPA ran hypothetical economic impact analyses and the results are in Tables 5-6 and 5-7.

5.5   Energy Impacts

Executive Order 13211 (66 FR 28355, May 22, 2001) provides that agencies will prepare and submit to the Administrator of the Office of Information and Regulatory Affairs, Office of Management and Budget, a Statement of Energy Effects for certain actions identified as "significant energy actions." Section 4(b) of Executive Order 13211 defines "significant energy actions" as any action by an agency (normally published in the *Federal Register*) that promulgates or is expected to lead to the promulgation of a final rule or regulation, including notices of inquiry, advance notices of proposed rulemaking, and notices of proposed rulemaking: (1) (i) that is a significant regulatory action under Executive Order 12866 or any successor order, and (ii) is likely to have a significant adverse effect on the supply, distribution, or use of energy;

or (2) that is designated by the Administrator of the Office of Information and Regulatory Affairs as a significant energy action.

This rule is not a significant energy action as designated by the Administrator of the Office of Information and Regulatory Affairs because it is not likely to have a significant adverse impact on the supply, distribution, or use of energy. In general, we expect the NSPS to improve technology,. By making the use of wood fuel less polluting and more efficient, we might see an increase in the use of wood fuel, which would relieve pressure on traditional coal- or petroleum-based energy sources. However, it is difficult to determine the precise energy impacts that might result from this rule because wood-fueled appliances compete with other biomass forms as well as more traditional oil, electricity, and natural gas. We have not determined the potential conversion to other types of fuels and their associated appliances if the consumer costs of wood-fueled appliances increase and at what level that increase would drive consumer choice.

5.6   Unfunded Mandates Reform Act (UMRA)

5.6.1   *Future and Disproportionate Costs*

The UMRA requires that we estimate, where accurate estimation is reasonably feasible, future compliance costs imposed by the rule and any disproportionate budgetary effects. Our estimates of the future compliance costs of the proposed rule are discussed previously in this RIA. The nationwide annualized average compliance cost of this proposed rule for directly affected appliances is $15.7 million in the 2014-2022 time frame (2010 dollars). Therefore, this proposed rule would not be subject to the requirements of Sections 202 or 205 of the UMRA.

This proposed rule would also not be subject to the requirements of Section 203 of UMRA because it contains no regulatory requirements that might significantly or uniquely affect small governments. The proposed rule would not apply to such governments and would impose no obligations upon them. We do not believe that there will be any disproportionate budgetary effects of the proposed rule on any particular areas of the country, state or local governments, types of communities (e.g., urban, rural), or particular industry segments.

5.6.2   *Effects on the National Economy*

The UMRA requires that we estimate the effect of the proposed rule on the national economy. To the extent feasible, we must estimate the effect on productivity, economic growth, full employment, creation of productive jobs, and international competitiveness of U.S. goods and services if we determine that accurate estimates are reasonably feasible and that such effect is relevant and material. The nationwide economic impact of the proposed rule is presented earlier in this RIA chapter. This analysis provides estimates of the effect of the proposed rule on most of the categories mentioned above, and these estimates are presented earlier in this RIA chapter. The nature of this rule is such that it is not practical for us to use existing approaches, such as the Morgenstern et al. approach,[55] to estimate the impact on employment to the regulated entities and others from this proposed rule. We explain why this is true, and provide impacts associated with the monitoring, recordkeeping, and reporting requirements to provide some understanding of what impacts this proposal will have on employment for affected firms in section 5.7 below.

5.6.3 *Executive Order 13045: Protection of Children from Environmental Health Risks and Safety Risks*

Executive Order 13045, "Protection of Children from Environmental Health Risks and Safety Risks" (62 FR 19885, April 23, 1997), applies to any rule that (1) is determined to be "economically significant," as defined under Executive Order 12866, and (2) concerns an environmental health or safety risk that EPA has reason to believe may have a disproportionate effect on children. If the regulatory action meets both criteria, EPA must evaluate the environmental health or safety effects of the planned rule on children and explain why the planned regulation is preferable to other potentially effective and reasonably feasible alternatives considered by the Agency.

This proposed rule is not subject to Executive Order 13045 (62 FR 19885, April 23, 1997) because the Agency does not believe the environmental health risks or safety risks addressed by this action present a disproportionate risk to children. The report, Analysis of Exposure to Residential Wood Combustion Emissions for Different Socio-Economic Groups, shows that on a nationwide basis, cancer risks due to residential wood smoke emissions among disadvantaged population groups generally are lower than the risks for the general population due to residential wood smoke emissions. One of the demographic variables examined for this report was that of children 18 years and younger.

This proposed rule is expected to reduce environmental impacts for everyone, including children. This action proposes emissions limits at the levels based on the best system of emissions reduction (BSER), as required by the Clean Air Act. Based on our analysis, we believe that this proposed rule would not have a disproportionate impact on children.

---

[55] Morgenstern, R. D., W. A. Pizer, and J. S. Shih. 2002. "Jobs versus the Environment: An Industry-Level Perspective." *Journal of Environmental Economics and Management* 43(3):412-436.

The public is invited to submit comments or identify peer-reviewed studies and data that assess effects of early-life exposure to smoke from residential wood heaters.

5.6.4 *Executive Order 12898: Federal Actions to Address Environmental Justice in Minority Populations and Low-Income Populations*

Executive Order 12898 (59 FR 7629 (Feb. 16, 1994)) establishes federal executive policy on environmental justice. Its main provision directs federal agencies, to the greatest extent practicable and permitted by law, to make environmental justice part of their mission by identifying and addressing, as appropriate, disproportionately high and adverse human health or environmental effects of their programs, policies, and activities on minority populations and low-income populations in the United States.

EPA has determined that this proposed rule would not have disproportionately high and adverse human health or environmental effects on minority, low-income or indigenous populations because it increases the level of environmental protection for all affected populations without having any disproportionately high and adverse human health or environmental effects on any population, including any minority, low-income or indigenous population. This proposed rule would establish national standards that would reduce primarily PM emissions from new residential wood heaters and, thus, would decrease the level of emissions to which all affected populations are exposed. The EPA defines "Environmental Justice" to include meaning involvement of all people regardless of race, color, national origin, or income with respect to the development, implementation, and enforcement of environmental laws, regulations, and polices. The EPA maintains an ongoing commitment to ensure environmental justice for all people, regardless of race, color, national origin, or income. Ensuring environmental justice means not only protecting human health and the environment for everyone, but also ensuring that all people are treated fairly and are given the opportunity to participate meaningfully in the development, implementation, and enforcement of environmental laws, regulations, and policies.

5.7   Employment Impacts

In addition to addressing the costs and benefits of the proposed rule, EPA has analyzed the impacts of this rulemaking on employment, which are presented in this section. While a standalone analysis of employment impacts is not included in a standard cost-benefit analysis, such an analysis is of particular concern in the current economic climate of sustained high unemployment. Executive Order 13563, states, "Our regulatory system must protect public health, welfare, safety, and our environment while promoting economic growth, innovation, competitiveness, and job creation" (emphasis added). . A discussion of labor requirements associated with the installation, operation, and maintenance of control requirements, as well as reporting and recordkeeping requirements is included in Section 5.1.6, on compliance costs, of this RIA. However, due to data and methodology limitations, we have not quantified the rule's effects on labor, or the effects induced by changes in workers' incomes. What follows is an overview of the various ways that environmental regulation can affect employment. EPA continues to explore the relevant theoretical and empirical literature and to seek public comments in order to ensure that the way EPA characterizes the employment effects of its regulations is valid and informative.

This proposed regulation is expected to affect employment in the United States through the regulated sector – residential wood heater manufacturers – and related sectors, specifically, masonry contractors and residential construction (e.g. performing masonry and other on-site fireplace construction), wholesalers and distributors, and retailers (e.g. home furnishing stores that sell wood heaters), and suppliers of substitutes for residential wood-burning heaters (e.g. electric or natural gas heaters). The production of devices like wood stoves, hydronic heaters, and fireplace inserts is included under the heating equipment category (NAICS 333414). The U.S. Census Bureau reports that, in 2011, the industry employed 15,925 workers (see Table 3-1 in Section 3.1.3 of this RIA). Based on company data obtained for this profile, the residential wood heaters industry has a large number of producers, and we were able to identify 635 firms, employing approximately 17,000 workers annually. Previous analysis suggests that the industry relies on seasonal labor, ramping up production in months leading up to winter and reducing employment and production during the warmer parts of the year (AEI, 1986).

As described in Section 3.2.3 of this RIA, demand for residential wood heaters has been declining steadily, as shown from 1989 to 2005, but has stabilized more recently. More households rely on wood fuel as a supplemental heat source rather than as a primary source. In 2010, 2.1% of total occupied homes in the United States relied on wood heat as the primary fuel source for home heating. About 10–12% of American households rely on wood when secondary wood heat demand is counted, according to the U.S. Census Bureau and the Energy Information Administration (EIA). Demand varies regionally, in part, due to availability of energy sources. Current regional demand patterns are expected to continue, with the Northeast and Northwest regions of the country driving wood fuel combustion demand, but analysts anticipate that the wood heat product market will be embraced in other areas of the country in which wood and biomass are viable and inexpensive fuel sources (Frost & Sullivan, 2010).

The extent to which an increase in the price of residential wood heaters due to this rule would reduce the sales depends on the elasticity of demand for residential wood heaters. However, there are no recent empirical estimates of the price elasticity of demand for residential wood heaters. An estimate of -1.6 was derived for use in the RIA for the current Residential Wood Combustion NSPS (EPA, 1986). Available estimates for residential energy and heating fuel demand generally are relatively inelastic (i.e., there are only very small changes in demand in response to an increase in energy or fuel prices). A recent RAND report suggests that in the short term, demand for electricity and natural gas in residential markets is relatively inelastic (Bernstein and Griffin, 2005). There are a number of close substitutes for residential wood heating devices that include electric and gas furnaces and space heaters. The extent to which consumers are able to substitute between these options is likely to vary depending on geographic location. Overall, the presence of good substitutes will increase the elasticity of demand for wood heating equipment. In contrast, if locally-available alternative heating fuels (e.g. electricity, fuel oil) are relatively higher-priced, it may make switching away from wood heating equipment less likely and, ultimately, make demand for wood heating equipment inelastic. Also, the elasticity may depend on whether the fuel in question is a secondary source of fuel instead of a primary fuel source. Based on the available information, including the RAND report, we do not expect sales of residential wood to fall substantially due to this rule, particularly in the near-term.

From an economic perspective labor is an input into producing goods and services; if a regulation requires that more labor be used to produce a given amount of output, that additional labor is reflected in an increase in the cost of production. Moreover, when the economy is at full employment, we would not expect an environmental regulation to have an impact on overall employment because labor is being shifted from one sector to another. On the other hand, in periods of high unemployment, employment effects (both positive and negative) are possible. For example, an increase in labor demand due to regulation may result in a short-term net increase in overall employment as workers are hired by the regulated sector to help meet new requirements (e.g., to install new equipment) or by the environmental protection sector to produce new abatement capital resulting in hiring previously unemployed workers . When significant numbers of workers are unemployed, the opportunity costs associated with displacing jobs in other sectors are likely to be smaller. And, in general, if a regulation imposes high costs and does not increase the demand for labor, it may lead to a decrease in employment. The responsiveness of industry labor demand depends on how these forces all interact. Economic theory indicates that the responsiveness of industry labor demand depends on a number of factors: price elasticity of demand for the product, substitutability of other factors of production, elasticity of supply of other factors of production, and labor's share of total production costs.

Berman and Bui (2001) put this theory in the context of environmental regulation, and suggest that, for example, if all firms in the industry are faced with the same compliance costs of regulation and product demand is inelastic, then industry output may not change much at all.

Regulations set in motion new orders for pollution control equipment and services. New categories of employment have been created in the process of implementing environmental regulations. When a regulation is promulgated, one typical response of industry is to order pollution control equipment and services in order to comply with the regulation when it becomes effective. On the other hand, the closure of plants that choose not to comply – and any changes in production levels at plants choosing to comply and remain in operation - occur after the compliance date, or earlier in anticipation of the compliance obligation. Environmental regulation may increase revenue and employment in the environmental technology industry. While these increases represent gains for that industry, they translate into costs to the regulated industries required to install the equipment.

Environmental regulations support employment in many basic industries. Regulated firms either hire workers to design and build pollution controls directly or purchase pollution control devices from a third party for installation. Once the equipment is installed, regulated firms hire workers to operate and maintain the pollution control equipment—much like they hire workers to produce more output In addition to the increase in employment in the environmental protection industry (via increased orders for pollution control equipment), environmental regulations also support employment in industries that provide intermediate goods to the environmental protection industry. The equipment manufacturers, in turn, order steel, tanks, vessels, blowers, pumps, and chemicals to manufacture and install the equipment. Currently in most cases there is no scientifically defensible way to generate sufficiently reliable estimates of the employment impacts in these intermediate goods sectors.

5.7.1 Employment Impacts within the Regulated Industry

It is sometimes claimed that new or more stringent environmental regulations raise production costs thereby reducing production which in turn must lead to lower employment. However, the peer-reviewed literature indicates that determining the direction of net employment effects in a regulated industry is challenging due to competing effects. Environmental regulations are assumed to raise production costs and thereby the cost of output, so we expect the "output" effect of environmental regulation to be negative (higher prices lead to lower sales). On the other hand, complying with the new or more stringent regulation requires additional inputs, including labor, and may alter the relative proportions of labor and capital used by regulated firms in their production processes. Two sets of researchers discussed here, Berman and Bui (2001) and Morgenstern, Pizer, and Shih (2002),[56] demonstrate using standard neoclassical microeconomics that environmental regulations have an ambiguous effect on employment in the regulated sector.[59] These theoretical results imply that the effect of environmental regulation on employment in the regulated sector is an empirical question and both sets of authors tested their models empirically using different methodologies. Both Berman and Bui and Morgenstern et al. examine the effect of environmental regulations on employment and both find that overall they had no significant net impact on employment in the sectors they examined.

Berman and Bui (2001) examine how an increase in local air quality regulation that reduces NOx emissions affects manufacturing employment in the South Coast Air Quality Management District (SCAQMD), which incorporates Los Angeles and its suburbs. During the time frame of their study, 1979 to 1992, the SCAQMD enacted some of the country's most stringent air quality regulations, which were more stringent than federal and state regulations. Using SCAQMD's local air quality regulations, Berman and Bui identify the effect of environmental regulations on net employment in the regulated industries.[57,58] The authors find that "while regulations do impose large costs, they have a limited effect on employment" (Berman and Bui, 2001, p. 269). Their conclusion is that local air quality regulation "probably increased labor demand slightly" but that "the employment effects of both compliance and increased stringency are *fairly precisely estimated zeros* [emphasis added], even when exit and dissuaded entry effects are included" (Berman and Bui, 2001, p. 269).[59]

Morgenstern et al. (2002) estimated the effects of pollution abatement expenditures on net employment in four highly regulated sectors (pulp and paper, plastics, steel, and petroleum refining). They conclude that increased abatement expenditures generally have *not* caused a significant change in net employment in those sectors. While the specific sectors Morgenstern et

---

[56] Berman, E. and L. T. M. Bui (2001). "Environmental Regulation and Labor Demand: Evidence from the South Coast Air Basin." Journal of Public Economics 79(2): 265-295.
Morgenstern, R. D., W. A. Pizer, and J. S. Shih. 2002. Jobs versus the Environment: An Industry-Level Perspective.|| Journal of Environmental Economics and Management 43(3):412-436.

[57] Note, like Morgenstern, Pizer, and Shih (2002), this study does not estimate the number of jobs created in the environmental protection sector.
[58] Berman and Bui include over 40 4-digit SIC industries in their sample.
[59] Including the employment effect of exiting plants and plants dissuaded from opening will increase the estimated impact of regulation on employment.

al. examined are different than the sectors considered here, the methodology that Morgenstern et al. developed is still an informative way to qualitatively assess the effects of this rulemaking on employment in the regulated sector.

While there is an extensive empirical, peer-reviewed literature analyzing the effect of environmental regulations on various economic outcomes including productivity, investment, competitiveness as well as environmental performance, there are only a few papers that examine the impact of environmental regulation on employment, but this area of the literature has been growing. As stated previously in this RIA section, empirical results from Berman and Bui (2001) and Morgenstern et al (2002) suggest that new or more stringent environmental regulations do not have a substantial impact on net employment (either negative or positive) in the regulated sector. Nevertheless, other empirical research suggests that more highly regulated counties may generate fewer jobs than less regulated ones (Greenstone 2002, Walker 2011). However, the methodology used in these two studies cannot estimate whether aggregate employment is lower or higher due to more stringent environmental regulation, it can only imply that relative employment growth in some sectors differs between more and less regulated areas. List et al. (2003) find some evidence that this type of geographic relocation, from more regulated areas to less regulated areas may be occurring. Overall, the peer-reviewed literature does not contain evidence that environmental regulation has a large impact on net employment (either negative or positive) in the long run across the whole economy.

While the theoretical framework laid out by Berman and Bui (2001) and Morgenstern et al. (2002) still holds for the industries affected under this proposed NSPS, important differences in the markets and regulatory settings analyzed in their study and the setting presented here lead us to conclude that it is inappropriate to utilize their quantitative estimates to estimate the employment impacts from this proposed regulation. In particular, the industries used in these two studies as well as the timeframe (late 1970's to early 1990's) are quite different than those in this proposed rule. Furthermore, the control strategies analyzed for this RIA mostly include process and design changes to reduce emissions during the production of affected heaters, and not after these heaters are in operation.[60] For instance, use of a catalyst combustor is common in wood stoves in order to reduce emissions and also improve heat efficiency. Retrofits are uncommon because replacing the wood stove is often a more economical alternative. On the other hand, the pollution control strategies examined by Berman and Bui and Morgenstern et al. are primarily

_____

[60] More detail on how emission reductions expected from compliance with this rule can be obtained can be found in Section 4 of this RIA.

add-on or end-of-line pollution controls. For these reasons we conclude there are too many uncertainties as to the transferability of the quantitative estimates in these two studies to apply their estimates to quantify the employment impacts within the regulated sectors for this proposed regulation.

Greenstone, M. (2002). "The Impacts of Environmental Regulations on Industrial Activity: Evidence from the 1970 and 1977 Clean Air Act Amendments and the Census of Manufactures." *Journal of Political Economy* 110(6): 1175-1219.

List, J. A., D. L. Millimet, P. G. Fredriksson, and W. W. McHone (2003). "Effects of Environmental Regulations on Manufacturing Plant Births: Evidence from a Propensity Score Matching Estimator." *The Review of Economics and Statistics* 85(4): 944-952.

Walker, Reed. (2011)."Environmental Regulation and Labor Reallocation." American Economic Review: Papers and Proceedings, 101(2).

SECTION 6
SMALL ENTITY SCREENING ANALYSIS

The Regulatory Flexibility Act as amended by the Small Business Regulatory Enforcement Fairness Act (SBREFA) generally requires an agency to prepare a regulatory flexibility analysis of any rule subject to notice and comment rulemaking requirements under the Administrative Procedure Act or any other statute, unless the agency certifies that the rule will not have a significant economic impact on a substantial number of small entities. Small entities include small businesses, small governmental jurisdictions, and small not-for-profit enterprises.

After considering the economic impact of the proposed rule on small entities, the screening analysis indicates that we cannot conclude that this proposed rule may not have a significant economic impact on a substantial number of small entities (or "SISNOSE") for certain residential wood heating products covered under the revised NSPS proposal. For this analysis EPA considered sales and revenue tests for establishments owned by representative small entities that manufacture or construct residential wood heating devices.

6.1   Small Entity Data Set

The industry sectors covered by the proposed rule were identified during the development of the cost analysis (see Sections 3 and 5). The Statistics of U.S. Businesses (SUSB) provides national information on the distribution of economic variables by industry and enterprise size (U.S. Census, 2008a, 2008b). The Census Bureau and the Office of Advocacy of the Small Business Administration (SBA) supported and developed these files for use in a broad range of economic analyses.[61] Statistics include the total number of establishments and receipts for all entities in an industry; however, many of these entities may not necessarily be covered by the final rule. SUSB also provides statistics by enterprise employment and receipt size.

The Census Bureau's definitions used in the SUSB are as follows:

- *Establishment*: An establishment is a single physical location where business is conducted or where services or industrial operations are performed.

- *Receipts*: Receipts (net of taxes) are defined as the revenue for goods produced, distributed, or services provided, including revenue earned from premiums, commissions and fees, rents, interest, dividends, and royalties. Receipts exclude all revenue collected for local, state, and federal taxes.

---

[61] See http://www.census.gov/csd/susb/ and http://www.sba.gov/advo/research/data.html for additional details.

- *Enterprise*: An enterprise is a business organization consisting of one or more domestic establishments that were specified under common ownership or control. The enterprise and the establishment are the same for single-establishment firms. Each multi-establishment company forms one enterprise—the enterprise employment and annual payroll are summed from the associated establishments. Enterprise size designations are determined by the summed employment of all associated establishments.

Because the SBA's business size definitions (SBA, 2013) apply to an establishment's "ultimate parent company," we assumed in this analysis that the "enterprise" definition above is consistent with the concept of ultimate parent company that is typically used for SBREFA screening analyses and the terms are used interchangeably.

6.2   Small Entity Economic Impact Measures

The analysis generated a set of establishment sales tests (represented as cost-to-receipt ratios) for NAICS codes associated with sectors listed in Table 6-1. Although the appropriate SBA size definition should be applied at the parent company (enterprise) level, we can only compute and compare ratios for a model establishment owned by an enterprise within an SUSB size range (employment or receipts). Using the SUSB size range helps us account for receipt differences between establishments owned by large and small enterprises and also allows us to consider the variation in small business definitions across affected industries. Using establishment receipts is also a somewhat conservative approach, because an establishment's parent company (the "enterprise") may have other economic resources that could be used to cover the costs of the final rule. It should be noted that these impacts are for the 2014–2022 time frame; as mentioned earlier in this RIA, the annualized costs considered in this analysis reflect an average of the proposal's compliance costs incurred by affected sources each year from promulgation through 2022.

6.2.1   Establishment Employment and Receipts

The sales test compares a representative establishment's total annual compliance costs to the average establishment receipts for enterprises in several size categories.[62] For industries with SBA employment size standards, we calculated average establishment receipts for each enterprise employment range (Table 6-2). For industries with SBA receipt size standards, we calculated average establishment receipts for each enterprise receipt range (Table 6-3). The analysis assumes that the majority of affected entities are covered under hardware manufacturing

---

[62] For the 1 to 20 employee category, we excluded SUSB data for enterprises with zero employees. These enterprises did not operate the entire year.

(NAICS 332510) and heating equipment manufacturing (NAICS 333414). We use establishment data from the 2007 Economic Census because that data is the most recent release of public establishment-level information for the industries impacted by this proposal.

Table 6-1.   Revised NSPS Proposal for Residential Wood Heating Devices: Affected Sectors and SBA Small Business Size Standards

Industry Description	Corresponding NAICS	SBA Size Standard for Businesses (July 22, 2013)	Type of Small Entity
New single-family general contractors	236115	$33.5 million in annual receipts	Masonry
Masonry contractors	238140	$14.0 million in annual receipts	Masonry
Hardware manufacturing	332510	500 employees	All product types
Heating equipment (except warm air) manufacturing	333414	500 employees	All product types
Plumbing and heating equipment and supplies (hydronics) merchant wholesalers	423720	100 employees	All product types
All other home furnishing stores	442299	$19.0 million in annual receipts	Business

However, the revised NSPS proposal has the potential to affect small entities classified as new home construction and masonry contractors. In addition, wholesalers of imported residential heating devices may also be affected if these establishments are required to certify imported products.

6.2.2   Establishment Compliance Cost

Annual entity compliance costs vary depending on the product type manufactured and the number of product models they would need to redesign under the revised NSPS proposal as mentioned in Section 5 of this RIA. For this analysis compliance costs were estimated based on the average development costs defined in the engineering cost analysis, presented in Section 5-1. The analysis assumes that manufacturers have between two and seven model fireboxes that would be subject to the new NSPS. There is limited information on the actual number of model fireboxes associated with small businesses. Hence, for purposes of the small entity screening analysis, we assumed that smaller companies maintain fewer than three firebox models that would be subject to the revised NSPS. In the absence of better data, EPA believes that between one and three firebox models is a reasonable assumption for our analysis of impacts to potentially affected small businesses.

Table 6-2. Average Receipts for Affected Industry by Enterprise Employment Size: 2007 ($2010 million/establishment)

NAICS	NAICS Description	SBA Size Standard for Businesses (effective July 22, 2013)	Owned by Enterprises with Employment Ranges:							
			All Enterprises	Fewer than 20 Employees	22 to 99 Employees	100 to 499 Employees	500 to 749 Employees	750 to 999 Employees	1,000 to 1,500 Employees	1,500+ Employees
332510	Hardware manufacturing	500 employees	$13.27	$1.33	$7.71	$26.26	$72.83	$62.18	$37.90	$62.27
333414	Heating equipment manufacturing	500 employees	$13.78	$1.27	$11.17	$39.81	NA	NA	NA	$8.09
423720	Plumbing and heating equipment wholesalers	100 employees	$7.49	$2.58	$9.68	$10.79	$9.82	NA	NA	$0.14

NA = Not available. SUSB did not report this data due to concern with disclosure of confidential information or other reasons. Escalation of average receipts from 2007 to 2010 is accomplished by use of the annual GDP implicit price deflator (available at http://research.stlouisfed.org/fred2/series/GDPDEF/downloaddata?cid=21). The escalation ratio between 2007 and 2010 is 1.045.

Table 6-3. Average Receipts for Affected Industry by Enterprise Receipt Range: 2007 ($2010 million/establishment)

NAICS	NAICS Description	SBA Size Standard for Businesses (effective July 22, 2013)	Owned By Enterprises with Receipt Range:								
			All Enterprises	Less than 100K Receipts	100 to 499K Receipts	500 to 999K Receipts	1,000 to 4,999K Receipts	5,000 to 9,999K Receipts	10,000 to 49,999K Receipts	50,000 to 99,999K Receipts	100,000K + Receipts
236115	New single-family general contractors	$33.5 million in annual receipts	$1.77	$0.05	$0.29	$0.77	$2.24	$7.907	$19.34	$59.14	$269.93
238140	Masonry Contractors	$14.0 million in annual receipts	$1.08	$0.05	$0.27	$0.77	$2.232	$7.37	$18.64	$48.03	$59.26
442299	All other home furnishing stores	$19.0 million in annual receipts	$2.68	$0.05	$0.28	$0.78	$1.99	$6.14	NA	NA	$401.23

NA = Not available. SUSB did not report this data due to concern with disclosure of confidential information or other reasons. Escalation of average receipts from 2007 to 2010 is accomplished by use of the annual GDP implicit price deflator (available at http://research.stlouisfed.org/fred2/series/GDPDEF/downloaddata?cid=21). The escalation ratio between 2007 and 2010 is 1.045.

Then, we computed per-entity compliance costs for representative establishments and for manufacturing each product type (see Table 6-4). For this analysis, the annualized costs as presented in Table 6-4 assumed the total model development costs for four model fireboxes spread over a 6-year model development time frame and scaled to a single model. Table 6-4 shows the estimated average annualized cost of $60,000 per model and its use in deriving the national total compliance costs for the proposed option. Table 6-5 presents the same costs for the alternative option. This cost was assumed to be constant for most product types. Lower compliance cost for pellet stoves due to the fact that most existing models already comply with the regulation. Lower bound on compliance cost for masonry heaters consists of a nominal licensing fee ($200) for the use of computer simulation model software to certify the site built units.

Table 6-4. Per-Entity Annualized Compliance Costs by Product Type—Proposed Option ($2010 millions)

Product Type	No. Establishments	Assumed Affected Models per Establishment[a]	Annual Compliance Cost per Model Firebox ($ millions)	Total Industry Costs—Proposed Option ($ millions)
Wood stoves	34	3	$0.04	$4.21
Single burn rate stoves	3	7	$0.04	$0.90
Pellet stoves	29	4	$0.03	$3.46
Forced-air furnaces	7	7	$0.05	$2.25
Masonry heaters[b]	48	2–8	< $0.001 to $0.003	$0.31
Hydronic heating systems	30	4	$0.04	$4.55
National Compliance Cost				**$15.7**

[a] Table totals may differ because of rounding.

[b] Masonry heater establishments include 2 large and 4 medium manufacturers, and 42 small custom builders.

For each case in this analysis, the number of models each representative establishment must redesign to comply with the NSPS emission limits in the options analyzed in this RIA varies by product type. The total annualized compliance cost per establishment is calculated by multiplying the number of firebox models requiring redesign by the annualized cost per model ($63,850). Table 6-6 presents the assumed number of models per establishment by product type. Figures 6-1 and 6-2 illustrate the distribution of compliances costs by product type.

Table 6-5. **Per-Entity Annualized Compliance Costs by Product Type—Alternative Option ($2010 millions)**

Product Type	No. Establishments	Assumed Affected Models per Establishment[a]	Annual Compliance Cost per Model Firebox ($ millions)	Total Industry Costs—Proposed Option ($ millions)
Wood stoves	34	3	$0.08	$8.09
Single burn rate stoves	3	7	$0.07	$1.54
Pellet stoves	29	4	$0.05	$6.25
Forced-air furnaces	7	7	$0.08	$3.81
Masonry heaters[b]	48	2–8	< $0.001 to $0.003	$0.31
Hydronic heating systems	30	4	$0.08	$8.30
National Compliance Cost				**$28.3**

[a] Table totals may differ because of rounding.

[b] Masonry heater establishments include 2 large and 4 medium manufacturers, and 42 small custom builders.

Table 6-6. **Representative Establishment Costs Used for Small Entity Analysis ($2010)**

	Best Estimate
Number of models requiring redesign	2
Annual cost per model	$63,850
Average annual cost per establishment	$127,700

For the sales test, we divided the representative establishment compliance costs reported in Table 6-6 by the representative establishment receipts reported in Tables 6-2 and 6-3. This is known as the cost-to-receipt (i.e., sales) ratio, or the "sales test." The "sales test" is the impact methodology EPA employs in analyzing small entity impacts as opposed to a "profits test," in which annualized compliance costs are calculated as a share of profits.

Information on annual revenues or sales is more commonly available data for entities normally affected by EPA regulations, and profits data normally made available are often not the true profit earned by firms because of accounting and tax considerations. Revenues as typically published are usually correct figures and are more reliably reported when compared with profit data. The use of a "sales test" for estimating small business impacts for a rulemaking such as this

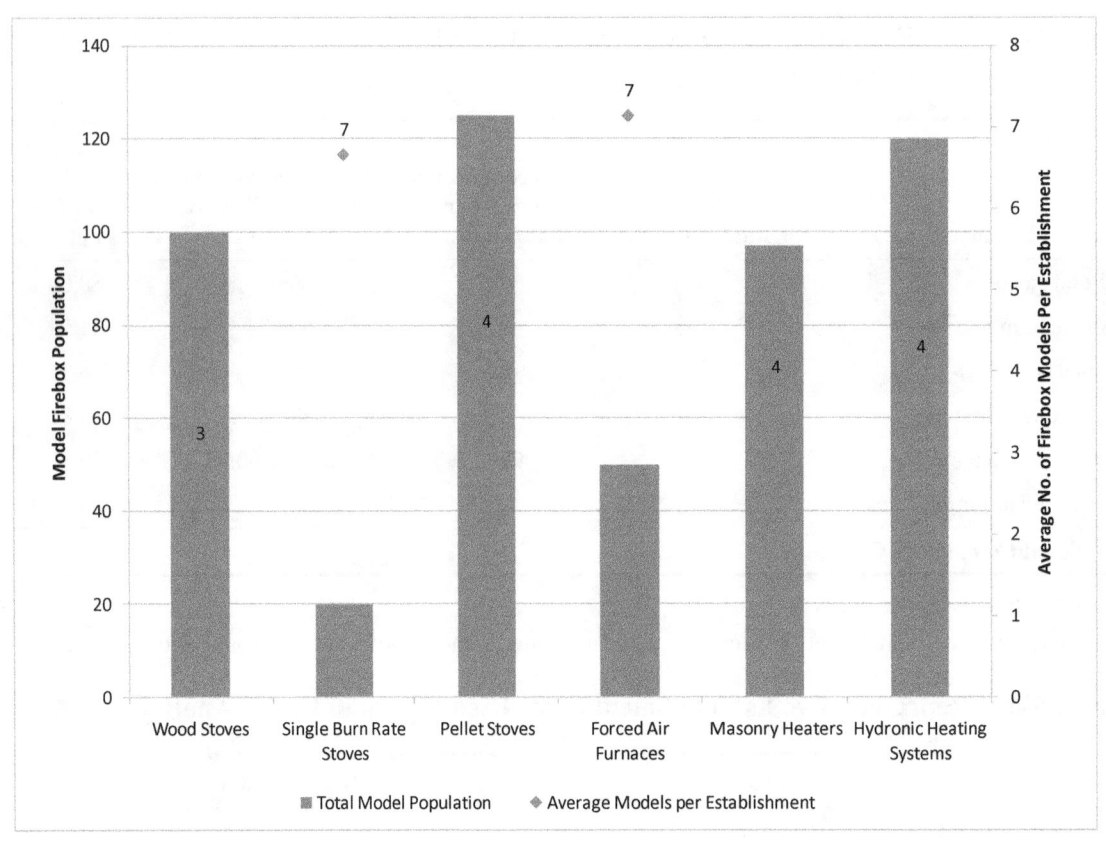

Figure 6-1.   Population of Firebox Models and Average Models per Establishment by Product Type

one is consistent with guidance offered by EPA on compliance with SBREFA[63] and is consistent with guidance published by the SBA's Office of Advocacy that suggests that cost as a percentage of total revenues is a metric for evaluating cost increases on small entities in relation to increases on large entities (SBA, 2003).[64] The annualized cost per sales for a company represents the maximum price increase in affected product needed for the company to completely recover the annualized costs imposed by the regulation.

---

[63] The SBREFA compliance guidance to EPA rule writers (EPA, 2006a) regarding the types of small business analysis that should be considered can be found at http://www.epa.gov/sbrefa/documents/rfaguidance11-00-06.pdf, pp. 24-25.

[64] This compliance guide produced by SBA can be found at http://www.sba.gov/sites/default/files/rfaguide_0512_0.pdf.

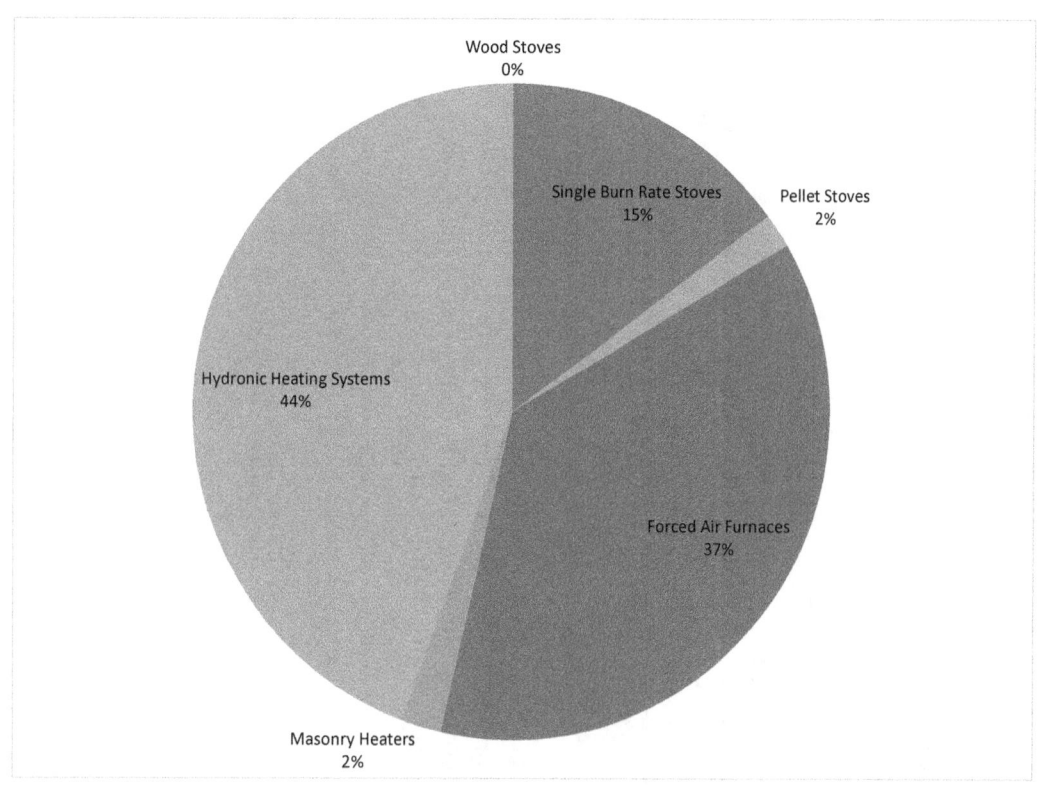

Figure 6-2.   Distribution of National Compliance Costs by Product Type in 2014 to 2022

For purposes of this analysis, EPA assumes most small entities in the residential wood heating industry are likely to manufacture fewer than three distinctive firebox models and in many cases they would support only one model. We assume for this analysis that most small entities in this industry will manufacture an average of two distinctive firebox models. Hence, EPA believes that the estimate in Table 6-6 above is the most representative establishment costs to assess impacts on small businesses. If the cost-to-receipt ratio is less than 1%, then we consider the rule to not have a significant impact on the establishment (and, company) in question. Table 6-7 presents the cost-to-receipt ratios for each category of establishments (establishments with ratios that exceed 1% under each case are highlighted) for the proposal option. Table 6-8 present these ratios for the alternative option.

Table 6-7.   Cost-to-Receipt Ratio Results for the Proposal Option by NAICS Code[a]

NAICS	Description	All Establishments	Fewer than 20 Employees	22 to 99 Employees	100 to 499 Employees
332510	Hardware manufacturing	3.12%	19.74%	3.40%	0.99%
333414	Heating equipment manufacturing	1.89%	20.73%	2.34%	0.67%
423720	Plumbing and heating equipment wholesalers	3.50%	10.15%	2.71%	2.43%

NAICS	Description	Total	Less than 100K Receipts	100 to 499K Receipts	500 to 999K Receipts	1,000 to 4,999K Receipts	5,000 to 9,999K Receipts	10,000 to 49,999K Receipts
236115	New single-family general contractors	7.18%	231.81%	43.78%	16.39%	5.62%	1.78%	0.69%
238140	Masonry contractors	11.59%	232.50%	47.70%	16.54%	5.43%	1.72%	0.74%
442299	All other home furnishing stores	7.86%	224.82%	46.00%	17.02%	7.47%	3.84%	

[a] All the cost to receipts results incorporate costs that are primarily R&D activities that firms will engage in to build appliance models that comply with the options analyzed in this RIA. The R&D cycle is estimated at 6 years, while the appliance life for all affected categories is 20 years.

Table 6-8. Cost-to-Receipt Ratio Results for the Alternative Option by NAICS Code[a]

NAICS	Description	All Establishments	Fewer than 20 Employees	22 to 99 Employees	100 to 499 Employees
332510	Hardware manufacturing	5.22%	33.05%	5.69%	1.66%
333414	Heating equipment manufacturing	3.17%	34.70%	3.92%	1.12%
423720	Plumbing and heating equipment wholesalers	5.87%	17.00%	4.54%	4.07%

NAICS	Description	Total	Less than 100K Receipts	100 to 499K Receipts	500 to 999K Receipts	1,000 to 4,999K Receipts	5,000 to 9,999K Receipts	10,000 to 49,999K Receipts
236115	New single-family general contractors	12.03%	389.44%	73.55%	27.54%	9.44%	2.99%	1.16%
238140	Masonry contractors	19.47%	390.60%	80.14%	27.79%	9.12%	2.89%	1.24%
442299	All other home furnishing stores	13.20%	377.70%	77.28%	27.23%	13.00%	6.45%	

[a] All the cost to receipts results incorporate costs that are primarily R&D activities that firms will engage in to build appliance models that comply with the options analyzed in this RIA. The R&D cycle is estimated at 6 years, while the appliance life for all affected categories is 20 years.

6.2.2.1 Analysis Results

In our small entity analysis for the proposed option, using an annual compliance cost of $127,000 as the estimated cost borne by affected small entities, establishments in NAICS 332510, 333414, and 423720 with fewer than 500 employees have cost-to-receipt ratios higher than 1%. Establishments in NAICS 236115, 238140, and 442299 with receipts less than $10 million have cost-to-receipt ratios higher than 1%. For the alternative option, the impacts are higher than 1% for all establishments in any size class.

After considering the economic impacts of this proposed rule on small entities, we cannot certify that this action will not have a significant economic impact on a substantial number of small entities. This certification is based on the economic impact of this action to all affected small entities across all industries affected. Using the estimate of impacts presented earlier in this chapter, we estimate that all small entities will have annualized costs of greater than 1% of their sales in all industries with fewer than 20 employees and NAICS 236115, 238140, and 442299 with receipts less than $10 million. Those establishments in NAICS 332510, 333414, and 423720 with cost-to-receipt ratios higher than 1% account for more than 80% of small entities. Establishments in NAICS 236115, 238140, and 442299 with cost-to-receipt ratios higher than 1% account for more than 99% of small entities. Small entity impacts are higher than this for the alternative option. We thus conclude that we cannot certify that there is not a significant economic impact on a substantial number of small entities (SISNOSE) for this rule.

It should be noted that the cost to receipts analysis included in this RIA reflect the large majority of annualized costs that are composed of research and development (R&D) activities (nearly 90%, based on the manufacturers' cost memorandum) that have a shorter life than the total life of affected appliances—six years for an R&D cycle as mentioned in section 5 of this RIA compared to 20 years for the life of affected appliances. The impacts on small entities should be understood in the context that a large share of the estimated annualized costs reflect expenses in the early years of the appliance life, and only a small share recurs each year over the entire appliance life of 20 years. In addition, the application of these costs for the options analyzed in this RIA will lead to a somewhat conservative (or, over-stated) cost estimate as stated previously in section 5.1.3. One example of this is that all hydronic heaters will undergo R&D beginning in 2013 to comply with either option analyzed; there is a small percentage of hydronic heaters that already meet the emission limits under each option considered in the RIA. Similar assumptions are also made in the cost estimates for single burn rate stoves and forced air furnaces. Given these two considerations, the costs that are included in the cost to sales analyses presented in this RIA are somewhat conservative in nature, and the cost to receipts impacts shown above should be understood in that context.

A sensitivity analysis showing the effect of the R&D cycle lifespan on the cost to sales estimates for each option is below. Table 6-9 shows how the model firebox costs that are input to the cost to sales (and small business) analyses will change with changes to the R&D cycle lifespan.

Table 6-9. Total Annual Cost (TAC) per Appliance Model – for Varying Annualized R&D Cycle Lifespans

Annualized R&D Cycle Lifespan (Years)	TAC/Appliance Model (2010$)	TAC per Establishment
6*	$63,850	$127,700
10	40,022	80,044
20	22,418	44,836

*As mentioned in Section 5 of this RIA, six years represents the annualized R&D cycle lifespan incorporated in the cost estimates for the Proposed RWH NSPS rule options.

The total annual cost per appliance model is the value of costs included in the cost to sales estimates to calculate these values. Each small entity is expected to modify 2 appliance models on average in order to comply with the proposal, an assumption we estimated earlier in this section of the RIA. We assume each small entity owns only one establishment (or place of business). Hence, the total annual cost to each small entity is twice the cost per appliance model. The annual cost to sales in the RIA will change proportionately to a change in the TAC since the sales estimates in the analysis will remain constant.

As shown in Table 6-9, with an increase in the R&D cycle lifespan from 6 to 10 years, the TAC/appliance model estimate falls to $40,022 from $63,850. The new TAC/appliance model estimate is 37% less than before. Thus, the cost to sales estimates will fall by 37% from the previous values. Finally, if the R&D cycle lifespan is reduced to 4 years, the TAC/appliance model estimate increases to $93,755 from $63,850. The new TAC/appliance model estimate is now 65% lower than before. Thus, the cost to sales estimates will decrease by 65% from the values that use an R&D cycle lifespan of 6 years.

Table 6-10 contains estimates of the changes in the cost to sales estimates for the Proposed option with an increase in the R&D cycle lifespan to 10 and 20 years from 6 years. The estimates with the change in R&D cycle lifespan to 10 and 20 years are in parentheses; the other values are those for the Proposed option.

Table 6-10. Cost-to-Sales Ratio Sensitivity Analysis Results Reflecting Different R&D Cycle Lifespans for the Proposed Option by NAICS Code*

NAICS	Industry Description	All Establishments	Establishments with Fewer than 20 Employees (%)	Establishments with Between 20 and 99 Employees (%)	Establishments with Between 100 and 499 Employees (%)
332510	Hardware manufacturing	3.12 (1.97, 1.09)	19.74 (12.44, 6.91)	3.40 (2.14, 1.19)	0.99 (0.62, 0.35)
333414	Heating equipment manufacturing	1.89 (1.19, 0.66)	20.73 (12.83, 7.26)	2.34(1.47, 0.82)	0.67 (0.42, 0.23)
423720	Plumbing and heating equipment wholesalers	3.50 (2.21, 1.23)	10.15 (6.39, 3.55)	2.71(1.71, 0.95)	2.43 (1.53, 0.85)

* The first value in parentheses is the cost to sales estimate for a 10 year R&D cycle lifespan; the second value is the cost to sales estimates for a 20 year R&D cycle lifespan.

6.3   Initial Regulatory Flexibility Analysis

An IRFA illustrates how EPA considers the proposed rule's small entity effects before a rule is finalized and provides information about how the objectives of the rule were achieved while minimizing significant economic impacts on small entities. We provide a summary of IRFA elements; the preamble for this rule provides additional details.

6.3.1   Reasons Why Action Is Being Considered

These proposals were developed following a Clean Air Act (CAA) section 111(b)(1)(B) periodic review of the existing residential wood heater new source performance standards (NSPS).

6.3.2   Statement of Objectives and Legal Basis of Proposed Rule

The EPA is proposing to amend Standards of Performance for New Residential Wood Heaters, and to add two new subparts: Standards of Performance for New Residential Hydronic Heaters and Forced-Air Furnaces and Standards of Performance for New Residential Masonry Heaters. These proposals are aimed at achieving several objectives, including applying tighter emission limits that reflect today's best systems of emission reduction (BSER); eliminating exemptions over a broad suite of residential wood combustion devices; revising test methods as appropriate; and streamlining the certification process. These proposals do not include any requirements on heaters that are solely fired by gas or oil. In addition, theses proposals do not include any requirements associated with wood heaters or other wood-burning appliances that are already in use. The EPA continues to encourage state, local, tribal, and consumer efforts to change out (replace) older heaters with newer, cleaner, more efficient heaters, but that is not part of this Federal rulemaking.

These proposals were developed following a Clean Air Act (CAA) section 111(b)(1)(B) periodic review of the existing residential wood heater new source performance standards (NSPS). The current body of evidence justifies that revision of the current residential wood heaters NSPS is needed to capture the improvements in performance of such units and to expand the applicability of this NSPS to include additional wood-burning residential heating devices. The changes being proposed with this action are aimed at achieving several objectives, including applying tighter emission limits that reflect today's best systems of emission reduction; eliminating exemptions over a broad suite of residential wood combustion devices; revising test methods as appropriate; and streamlining the certification process.

6.3.3 *Description and Estimate of the Number of Small Entities*

Small entities that EPA anticipates being affected by the standards would include almost all manufacturers of wood heaters listed in Section 2.2 of this document. EPA estimates that roughly 250–300 U.S. companies manufacture residential wood heaters. EPA believes that approximately 90% of these manufacturers meet the SBA small-entity definition of having fewer than 500 employees.

6.3.4 *Description and Compliance Costs*

A discussion of the methodology used to estimate cost impacts is presented in Section 5 of this RIA.

As required by section 609(b) of the RFA, as amended by SBREFA, EPA has conducted outreach to small entities and convened a SBAR Panel to obtain advice and recommendation of representatives of the small entities potentially subject to the requirements of this rule. On August 4, 2010, EPA's Small Business Advocacy Chairperson convened a Panel under section 609(b) of the RFA. In addition to the Chair, the Panel consisted of representatives of the Director of the Outreach and Information Division within EPA's Office of Air and Radiation, the Chief Counsel for Advocacy of the SBA, and the Administrator of the Office of Information and Regulatory Affairs within the Office of Management and Budget.

Based on consultations with the SBA, and resulting from solicited self-nominations, we prepared a list of 30 potential small entity representatives (SERs), from residential wood heating appliance manufacturers (wood stoves, pellet stoves, hydronic heaters, forced-air furnaces, and masonry heaters), other wood burning appliance manufacturers (fireplaces, cook stoves), equipment suppliers, chimney sweeps, test laboratories, masons, and trade associations. Once the pre-Panel process began and potential SERs were identified, EPA held an outreach meeting with the potential SERs and invited representatives from SBA's Office of Advocacy and the Office of Information and Regulatory Affairs within the Office of Management and Budget on June 29, 2010, to solicit their feedback on the upcoming proposed rulemaking. Representatives from 26 of the 30 companies and organizations that we selected as potential SERs for this SBREFA process participated in the meeting (in person and by phone). At that meeting EPA solicited written comments from the potential SERs, which were later summarized and shared with the Panel as part of the Panel convening document.

After the SBAR Panel was convened, the Panel distributed additional information to the SERs on August 11 and August 19, 2010, for their review and comment and in preparation for another outreach meeting. On August 25, 2010, the Panel met with the SERs to hear their comments on the information distributed via email. The Panel received written comments from the SERs in response to the discussions at this meeting and the outreach materials. The Panel asked the SERs to evaluate how they would be affected and to provide advice and recommendations regarding early ideas to provide flexibility.

Many of the SERs and the Panel had concerns about the breadth of the potential options discussed for this rulemaking and the challenges EPA would face in potentially conducting rulemaking for all of these source categories at one time and the challenges that the small businesses would face in having to potentially comply with standards for all of these source categories at one time. The Panel recommended that EPA should consider focusing efforts first on emissions sources that have the greatest potential to impact public health through the magnitude of emissions and population exposure. The EPA has narrowed the scope of this proposal to focus on the sources with the greatest potential impacts on public health. The Panel was sensitive to the need to carefully develop a rule that will minimize business closures, while still achieving significant emission reductions.

6.3.5 Panel Recommendations for Small Business Flexibilities

The Panel recommended that EPA consider and seek comment on an extensive range of regulatory alternatives to mitigate the impacts of the rulemaking on small businesses, including the options listed below. The following section summarizes the SBAR Panel recommendations.

Consistent with the RFA/SBREFA requirements, the Panel evaluated the assembled materials and comments related to elements of the IRFA. A copy of the Final Panel Report (including all comments received from SERs in response to the Panel's outreach meetings), as well as summaries of both outreach meetings that were held with the SERs, is included in the docket for the proposed rules. The following paragraphs are a subset of the full report.

The Panel encouraged EPA to consider flexibilities that will most directly minimize the small business burdens: Exemptions from the standards based on very low volume production, and delayed compliance dates for low volume production. The delayed compliance approach is predicated on the concept that it will take a number of years for manufacturers to recover the costs of the R&D investment in order to achieve compliance.

The Panel recommended that the EPA Administrator should consider the availability and feasibility of certification, testing labs, testing standards, and other requirements.

The Panel recommended that the EPA Administrator should consider emphasizing that the NSPS will address only new units, and the EPA Administrator should consider clarifying whether exemptions will be considered for historic replica equipment and historic property renovations.

EPA is looking at opportunities for reducing the burden on small entities of potential reporting, record keeping, and compliance requirements. For reporting and record keeping requirements in the revised NSPS, EPA is considering providing flexibilities similar to those in the 1988 NSPS. For example, the Panel recommended that EPA continue allowing manufacturers to keep records and report test results for a representative model appliance rather than testing and reporting results for each individual unit.

Many SERs expressed concern about potential compliance requirements associated with the planned proposed standards. Specifically, SERs anticipated potential logjams at third-party testing facilities as a result of EPA's regulating a broader range of product categories, which the SERs believe will slow down the certification process. In addition, many SERs are concerned about the costs associated with compliance requirements, including research and development, preliminary testing and certification of new products and recertification of products approved under the 1988 NSPS. The Panel recommended that EPA consider ways to streamline compliance certification, in particular, identifying flexible approaches and procedures that will reduce the burden and time for manufacturers to complete the application, testing and approval process for new model lines. For example, the Panel recommended that EPA consider allowing the use of International Standards Organization (ISO)-accredited laboratories and certifying bodies to expand the number of facilities that would be required for testing and certification of the new residential solid biomass combustion appliances. Additionally, the Panel recommended that EPA consider different compliance time frames for different product categories to reduce the potential for logjams at test labs and the overall impact on companies that manufacture multiple categories. Flexible compliance schedules would also help manufacturers of additional new appliances, such as hydronic heaters and forced-air furnaces, which were not subject to the 1988 standards.

Consistent with the RFA/SBREFA requirements, the Panel evaluated the assembled materials and small-entity comments on issues related to elements of the IRFA. A copy of the Panel report is included in the docket for this proposed rule. We invite comments on all aspects of the proposal and its impacts on small entities.

SECTION 7
HUMAN HEALTH BENEFITS OF EMISSIONS REDUCTIONS

7.1 Synopsis

Implementation of emissions limits required by the proposed residential wood heaters NSPS is expected to reduce direct emissions of $PM_{2.5}$. These reductions result from the imposition of tightened and new PM emissions limits for a number of emissions categories as described in Section 2 of this RIA. In this section, we quantify the monetized benefits for this rule associated with reduced exposure to ambient fine particulate matter ($PM_{2.5}$) resulting from the reduction of direct emissions of $PM_{2.5}$. The total $PM_{2.5}$ reductions are the consequence of the expected design changes to the affected appliances needed in order to meet the limits in the options analyzed in this RIA. We estimate the total monetized benefits for the proposed option to be $1.8 billion to $4.2 billion at a 3% discount rate and $1.7 billion to $3.8 billion at a 7% discount rate on a yearly average between 2014 and 2022. For the alternative option and same time frame, we estimate that the total monetized benefits are $1.9 billion to $4.2 billion at a 3% discount rate and $1.7 billion to $3.8 billion at a 7% discount rate. All estimates are in 2010$. These estimates reflect the monetized human health benefits of reducing cases of morbidity and premature mortality among populations exposed to $PM_{2.5}$ reduced by this rule.

Data, resources, and methodological limitations prevented EPA from monetizing the benefits from several important benefit categories. Included among the nonmonetized benefits are those associated with reduced exposure to about 3,200 tons of VOCs. VOCs are also precursors to ozone formation and therefore reducing health impact due to ozone exposure. Further, this rule would reduce each year 33,000 tons of CO, black carbon emissions, several HAP emissions such as benzene, formaldehyde, and dioxin. This rule will also reduce ecosystem effects, and visibility impairment due to PM emissions.

7.2 $PM_{2.5}$-Related Human Health Benefits

This rule is expected to reduce direct emissions of PM and emissions of VOCs, which are precursors to formation of ambient $PM_{2.5}$. Therefore, reducing these emissions would also reduce human exposure to ambient $PM_{2.5}$ and the incidence of $PM_{2.5}$-related health effects. In this section, we provide an overview of the $PM_{2.5}$-related benefits. A full description of the underlying data, studies, and assumptions is provided in the PM NAAQS RIA (U.S. EPA, 2012a).

In implementing this rule, emission controls may lead to reductions in ambient $PM_{2.5}$ concentrations below the National Ambient Air Quality Standards (NAAQS) for PM in some areas and assist other areas with attaining the PM NAAQS. Because the PM NAAQS RIA (U.S. EPA, 2012a) also calculated PM benefits, there are important differences worth noting in the design and analytical objectives of each RIA. The NAAQS RIAs illustrate the potential costs and benefits of attaining a revised air quality standard nationwide based on an array of emission reduction strategies for different sources including known and unknown controls, incremental to implementation of existing regulations and controls needed to attain the current standards. In short, NAAQS RIAs hypothesize, but do not predict, the reduction strategies that States may choose to enact when implementing a revised NAAQS. The setting of a NAAQS does not directly result in costs or benefits, and as such, the NAAQS RIAs are merely illustrative and the estimated costs and benefits are not intended to be added to the costs and benefits of other regulations that result in specific costs of control and emission reductions. However, it is possible that some costs and benefits associated with the required emission controls estimated in this RIA may account for the same air quality improvements as estimated in the illustrative PM NAAQS RIA.

By contrast, the emission reductions for implementation rules such as this rulemaking are generally for specific, well-characterized sources. In general, EPA is more confident in the magnitude and location of the emission reductions for implementation rules. As such, emission reductions achieved under these and other promulgated implementation rules will ultimately be reflected in the baseline of future NAAQS analyses, which would reduce the incremental costs and benefits associated with attaining revised future NAAQS. EPA remains forward looking towards the next iteration of the 5-year review cycle for the NAAQS. As a result, EPA does not re-issue NAAQS RIAs that retroactively update the baseline to account for implementation rules promulgated after a NAAQS RIA outside of the NAAQS review process. For more information on the relationship between the NAAQS and rules that are not ambient standards, such as analyzed here, please see Section 1.3 of the PM NAAQS RIA (U.S. EPA, 2012a).

7.2.1 Health Impact Assessment

The *Integrated Science Assessment for Particulate Matter* (PM ISA) (U.S. EPA, 2009) identified the human health effects associated with ambient $PM_{2.5}$, which include premature mortality and a variety of morbidity effects associated with acute and chronic exposures. Table 7-1 provides the quantified and unquantified benefits captured in EPA's benefits estimates for reduced exposure to ambient $PM_{2.5}$. Although the table below does not include entries for the unquantified health effects such as exposure to ozone and NO_2 nor welfare effects such as ecosystem effects and visibility impairment, these effects are itemized in Chapters 5 and 6 of the

Table 7-1. Human Health Effects of Ambient PM$_{2.5}$

Category	Specific Effect	Effect Has Been Quantified	Effect Has Been Monetized	More Information in PM NAAQS RIA
Improved Human Health				
Reduced incidence of premature mortality from exposure to PM$_{2.5}$	Adult premature mortality based on cohort study estimates and expert elicitation estimates (age >25 or age >30)	✓	✓	Section 5.6
	Infant mortality (age <1)	✓	✓	Section 5.6
Reduced incidence of morbidity from exposure to PM$_{2.5}$	Non-fatal heart attacks (age > 18)	✓	✓	Section 5.6
	Hospital admissions—respiratory (all ages)	✓	✓	Section 5.6
	Hospital admissions—cardiovascular (age >20)	✓	✓	Section 5.6
	Emergency room visits for asthma (all ages)	✓	✓	Section 5.6
	Acute bronchitis (age 8–12)	✓	✓	Section 5.6
	Lower respiratory symptoms (age 7–14)	✓	✓	Section 5.6
	Upper respiratory symptoms (asthmatics age 9–11)	✓	✓	Section 5.6
	Asthma exacerbation (asthmatics age 6–18)	✓	✓	Section 5.6
	Lost work days (age 18–65)	✓	✓	Section 5.6
	Minor restricted-activity days (age 18–65)	✓	✓	Section 5.6
	Chronic Bronchitis (age >26)	—[a]	—[a]	Section 5.6
	Emergency room visits for cardiovascular effects (all ages)	—[a]	—[a]	Section 5.6
	Strokes and cerebrovascular disease (age 50–79)	—[a]	—[a]	Section 5.6
	Other cardiovascular effects (e.g., other ages)	—	—	PM ISA[b]
	Other respiratory effects (e.g., pulmonary function, non-asthma ER visits, non-bronchitis chronic diseases, other ages and populations)	—	—	PM ISA[b]
	Reproductive and developmental effects (e.g., low birth weight, pre-term births, etc.)	—	—	PM ISA[b,c]
	Cancer, mutagenicity, and genotoxicity effects	—	—	PM ISA[b,c]

[a] We assess these benefits qualitatively due to time and resource limitations for this analysis. In the PM NAAQS RIA, these benefits were quantified in a sensitivity analysis, but not in the core analysis.

[b] We assess these benefits qualitatively because we do not have sufficient confidence in available data or methods.

[c] We assess these benefits qualitatively because current evidence is only suggestive of causality or there are other significant concerns over the strength of the association.

PM NAAQS RIA (U.S. EPA, 2012a). It is important to emphasize that the list of unquantified benefit categories is not exhaustive, nor is quantification of each effect complete.

We follow a "damage-function" approach in calculating benefits, which estimates changes in individual health endpoints (specific effects that can be associated with changes in air quality) and assigns values to those changes assuming independence of the values for those individual endpoints. Because EPA rarely has the time or resources to perform new research to measure directly either the health outcomes or their values for regulatory analyses, our estimates are based on the best available methods of benefits transfer, which is the science and art of adapting primary research from similar contexts to estimate benefits for the environmental quality change under analysis.

The health impact assessment (HIA) quantifies the changes in the incidence of adverse health impacts resulting from changes in human exposure to $PM_{2.5}$ or other air pollutants. We use the environmental *Ben*efits *M*apping and *A*nalysis *P*rogram (BenMAP) to systematize health impact analyses by applying a database of key input parameters, including population projections, health impact functions, valuation functions (Abt Associates, 2012). For this assessment, the HIA is limited to those health effects that are directly linked to ambient $PM_{2.5}$ concentrations. There may be other indirect health impacts associated with implementing emissions controls, such as occupational health exposures. Epidemiological studies generally provide estimates of the relative risks of a particular health effect for a given increment of air pollution (often per 10 $\mu g/m^3$ for $PM_{2.5}$). These relative risks can be used to develop risk coefficients that relate a unit reduction in $PM_{2.5}$ to changes in the incidence of a health effect. We refer the reader to section 5.6 of the PM NAAQS RIA for more information regarding the epidemiology studies and risk coefficients applied in this analysis (U.S. EPA, 2012a), and we briefly elaborate on adult premature mortality below. The size of the mortality effect estimates from epidemiological studies, the serious nature of the effect itself, and the high monetary value ascribed to prolonging life make mortality risk reduction the most significant health endpoint quantified in this analysis.

Considering a substantial body of published scientific literature, reflecting thousands of epidemiology, toxicology, and clinical studies, the PM ISA documents the association between elevated $PM_{2.5}$ concentrations and adverse health effects, including increased premature mortality (U.S. EPA, 2009). The PM ISA, which was twice reviewed by the Clean Air Scientific Advisory Committee of EPA's Science Advisory Board (SAB-CASAC) (U.S. EPA-SAB, 2009b, 2009c), concluded that there is a causal relationship between mortality and both long-term and short-term exposure to $PM_{2.5}$ based on the entire body of scientific evidence. The PM ISA also concluded that the scientific literature consistently finds that a no-threshold log-linear model most adequately portrays the PM-mortality concentration-response relationship while recognizing potential uncertainty about the exact shape of the concentration-response function.

For mortality, we use the effect coefficients from the most recent epidemiology studies examining two large population cohorts: the American Cancer Society (ACS) cohort (Krewski et al., 2009) and the Harvard Six Cities cohort (Lepeule et al., 2012). The PM ISA (U.S. EPA, 2009) concluded that the ACS and Six Cities cohorts provide the strongest evidence of the association between long-term $PM_{2.5}$ exposure and premature mortality with support from a number of additional cohort studies. The SAB's Health Effects Subcommittee (SAB-HES) also supported using these two cohorts for analyses of the benefits of PM reductions (U.S. EPA-SAB, 2010a). As both the ACS and Six Cities cohort studies have inherent strengths and weaknesses, we present benefits estimates using relative risk estimates from both these cohorts (Krewski et al., 2009; Lepeule et al., 2012).

As a characterization of uncertainty regarding the $PM_{2.5}$-mortality relationship, EPA graphically presents benefits derived from EPA's expert elicitation study (Roman et al., 2008; IEc, 2006). The primary goal of the 2006 study was to elicit from a sample of health experts probabilistic distributions describing uncertainty in estimates of the reduction in mortality among the adult U.S. population resulting from reductions in ambient annual average $PM_{2.5}$ levels. In that study, twelve experts provided independent opinions of the $PM_{2.5}$-mortality concentration-response function. Because the experts relied upon the ACS and Six Cities cohort studies to inform their concentration-response functions, the benefits estimates derived from the expert responses generally fall between results derived from the these studies (see Figure 7-1). We do not combine the expert results in order to preserve the breadth and diversity of opinion on the expert panel. This presentation of the expert-derived results is generally consistent with SAB advice (U.S. EPA-SAB, 2008), which recommended that the EPA emphasize that "scientific differences existed only with respect to the magnitude of the effect of $PM_{2.5}$ on mortality, not whether such an effect existed" and that the expert elicitation "supports the conclusion that the benefits of $PM_{2.5}$ control are very likely to be substantial." Although it is possible that newer scientific literature could revise the experts' quantitative responses if elicited again, we believe that these general conclusions are unlikely to change.

7.2.2 Economic Valuation

After quantifying the change in adverse health impacts, we estimate the economic value of these avoided impacts. Reductions in ambient concentrations of air pollution generally lower the risk of future adverse health effects by a small amount for a large population. Therefore, the appropriate economic measure is willingness to pay (WTP) for changes in risk of a health effect. For some health effects, such as hospital admissions, WTP estimates are generally not available, so we use the cost of treating or mitigating the effect. These cost-of-illness (COI) estimates generally (although not necessarily in every case) understate the true value of reductions in risk of a health effect. They tend to reflect the direct expenditures related to treatment but not the value of avoided pain and suffering from the health effect. The unit values applied in this analysis are provided in Table 5-9 of the PM NAAQS RIA for each health endpoint (U.S. EPA, 2012a).

Avoided premature deaths account for 98% of monetized PM-related benefits. The economics literature concerning the appropriate method for valuing reductions in premature mortality risk is still developing. The adoption of a value for the projected reduction in the risk of premature mortality is the subject of continuing discussion within the economics and public policy analysis community. Following the advice of the SAB's Environmental Economics Advisory Committee (SAB-EEAC), the EPA currently uses the value of statistical life (VSL) approach in calculating estimates of mortality benefits, because we believe this calculation provides the most reasonable single estimate of an individual's willingness to trade off money for reductions in mortality risk (U.S. EPA-SAB, 2000). The VSL approach is a summary measure for the value of small changes in mortality risk experienced by a large number of people.

EPA continues work to update its guidance on valuing mortality risk reductions, and the Agency consulted several times with the SAB-EEAC on the issue. Until updated guidance is available, the Agency determined that a single, peer-reviewed estimate applied consistently best reflects the SAB-EEAC advice it has received. Therefore, EPA has decided to apply the VSL that was vetted and endorsed by the SAB in the *Guidelines for Preparing Economic Analyses* (U.S. EPA, 2000)[65] while the Agency continues its efforts to update its guidance on this issue. This approach calculates a mean value across VSL estimates derived from 26 labor market and contingent valuation studies published between 1974 and 1991. The mean VSL across these studies is $6.3 million (2000$).[66]

---

[65] In the updated *Guidelines for Preparing Economic Analyses* (U.S. EPA, 2010e), EPA retained the VSL endorsed by the SAB with the understanding that further updates to the mortality risk valuation guidance would be forthcoming in the near future.

[66] In 1990$, this VSL is $4.8 million.

We then adjust this VSL to account for the currency year used in this RIA and to account for income growth from 1990 to the analysis year. The adjusted value for VSL is $8.0 million ($2010).

The Agency is committed to using scientifically sound, appropriately reviewed evidence in valuing mortality risk reductions and has made significant progress in responding to the SAB-EEAC's specific recommendations. In the process, the Agency has identified a number of important issues to be considered in updating its mortality risk valuation estimates. These are detailed in a white paper on "Valuing Mortality Risk Reductions in Environmental Policy," (U.S. EPA, 2010c) which recently underwent review by the SAB-EEAC. A meeting with the SAB on this paper was held on March 14, 2011 and formal recommendations were transmitted on July 29, 2011 (U.S. EPA-SAB, 2011). Draft guidance responding to SAB recommendations will be developed shortly.

In valuing premature mortality, we discount the value of premature mortality occurring in future years using rates of 3% and 7% (OMB, 2003). We assume that there is a "cessation" lag between changes in PM exposures and the total realization of changes in health effects. Although the structure of the lag is uncertain, the EPA follows the advice of the SAB-HES to assume a segmented lag structure characterized by 30% of mortality reductions in the first year, 50% over years 2 to 5, and 20% over the years 6 to 20 after the reduction in $PM_{2.5}$ (U.S. EPA-SAB, 2004c). Changes in the cessation lag assumptions do not change the total number of estimated deaths but rather the timing of those deaths.

7.2.3 *Benefit-per-ton Estimates*

Due to analytical limitations, it was not possible to conduct air quality modeling for this rule. Instead, we used a "benefit-per-ton" approach to estimate the benefits of this rulemaking. EPA has applied this approach in several previous RIAs (e.g., U.S. EPA, 2011b, 2011d, 2012b). These benefit-per-ton estimates provide the total monetized human health benefits (the sum of premature mortality and premature morbidity) of reducing one ton of $PM_{2.5}$ (or $PM_{2.5}$ precursor such as NO_x or SO_2) from a specified source. Specifically, in this analysis, we multiplied the estimates from the "Residential Wood Heaters" sector[67,68] by the corresponding emission reductions. The method used to derive these estimates is described in the Technical Support

---

[67] As explained in the TSD (U.S. EPA, 2013), we only have benefit-per-ton estimates for certain analysis years (i.e., 2005, 2016, 2020, 2025, and 2030). For this RIA, we selected the benefit-per-ton estimate closest to the analysis year for this RIA.

[68] Data from year 2020 was used as the year closest to the full implementation year for both options analyzed in this RIA—2019 for the Proposal option, 2022 for the Alternative option.

Document (TSD) on estimating the benefits-per-ton of reducing $PM_{2.5}$ and its precursors (U.S. EPA, 2013). One limitation of using the benefit-per-ton approach is an inability to provide estimates of the health benefits associated with exposure to HAP, CO, NO_2 or ozone.

The benefit-per-ton estimates described in the TSD (U.S. EPA, 2013) were derived using the approach published in Fann et al. (2012), but they have since been updated to reflect the studies and population data in the final PM NAAQS RIA (U.S. EPA, 2012a). The approach in Fann et al. (2012) is similar to the work previously published by Fann et al. (2009), but the newer study includes improvements that EPA believes would provide more reliable estimates of $PM_{2.5}$-related health benefits for emissions reductions in specific sectors. Specifically, the air quality modeling data reflect sectors that are more narrowly defined. In addition, the updated air quality modeling data reflect more recent emissions data (2005 rather than 2001) and has higher spatial resolution (12km rather than 36 km grid cells).

As noted below in the characterization of uncertainty, all benefit-per-ton estimates have inherent limitations. Specifically, all national-average benefit-per-ton estimates reflect the geographic distribution of the modeled emissions, which may not exactly match the emission reductions in this rulemaking, and they may not reflect local variability in population density, meteorology, exposure, baseline health incidence rates, or other local factors for any specific location.

Even though we assume that all fine particles have equivalent health effects, the benefit-per-ton estimates vary between precursors depending on the location and magnitude of their impact on $PM_{2.5}$ levels, which drive population exposure. The sector-specific modeling does not provide estimates of the $PM_{2.5}$-related benefits associated with reducing VOC emissions, but these unquantified benefits are generally small compared to other $PM_{2.5}$ precursors (U.S. EPA, 2012a).

7.2.4 *$PM_{2.5}$ Benefits Results*

Table 7-2 summarizes the monetized PM-related health benefits by precursor pollutant, including the emission reductions and benefit-per-ton estimates using discount rates of 3% and 7%. Benefits estimates are based on the average of annual emission reductions from proposed rule implementation between 2014 and 2022 (inclusive). Table 7-3 provides a summary of the reductions in health incidences associated with these pollution reductions. Figure 7-1 provides a visual representation of the range of $PM_{2.5}$-related benefits estimates using concentration-response functions from Krewski et al. (2009) and Lepeule et al. (2012) as well as

Table 7-2. Summary of Monetized $PM_{2.5}$-Related Health Benefits Estimates for the Proposed Residential Wood Heaters NSPS in the 2014–2022 Time Frame (2010$)[a]

Pollutant	Emissions Reductions (tons)	Benefit per ton (Krewski, 3%)	Benefit per ton (Lepeule, 3%)	Benefit per ton (Krewski, 7%)	Benefit per ton (Lepeule, 7%)	Total Monetized Benefits (millions 2010$ at 3%)		Total Monetized Benefits (millions 2010$ at 7%)	
Proposed									
Direct $PM_{2.5}$	4,825	$380,000	$860,000	$350,000	$780,000	$1,800	to $4,200	$1,700	to $3,800
$PM_{2.5}$ Precursors									
VOC[b]	3,250	—	—	—	—	—	to —	—	to —
				Total		$1,800	to $4,200	$1,700	to $3,800
Alternative									
Direct $PM_{2.5}$	4,878	$380000	$860,000	$350,000	$780,000	$1,900	to $4,200	$1,700	to $3,800
$PM_{2.5}$ Precursors									
VOC[b]	3,250	—	—	—	—	—	to —	—	to —
				Total		$1,900	to $4,200	$1,700	to $3,800

[a] All estimates reflect the average of annual emission reductions expected to occur between 2014 and 2022 (inclusive) resulting from proposed rule implementation. All estimates are rounded to two significant figures so numbers may not sum across columns. It is important to note that the monetized benefits do not include reduced health effects from direct exposure to NO_2, ozone exposure, ecosystem effects, or visibility impairment. All fine particles are assumed to have equivalent health effects, but the benefit per ton estimates vary depending on the location and magnitude of their impact on $PM_{2.5}$ levels, which drive population exposure. The monetized benefits incorporate the conversion from precursor emissions to ambient fine particles. Confidence intervals are unavailable for this analysis because of the benefit-per-ton methodology.

[b] Estimates of VOCs health benefits are currently not monetized and will be addressed only qualitatively.

Table 7-3.   Summary of Reductions in Health Incidences from PM$_{2.5}$-Related Benefits for the Proposed Residential Wood Heaters NSPS in the 2014-2022 Time Frame[a]

Avoided Premature Mortality	Proposal	Alternative
Krewski et al. (2009) (adult)	210	210
Lepeule et al. (2012) (adult)	470	480
Avoided Morbidity		
Emergency department visits for asthma (all ages)	110	100
Acute bronchitis (age 8–12)	320	320
Lower respiratory symptoms (age 7–14)	4,100	4,200
Upper respiratory symptoms (asthmatics age 9–11)	5,900	6,000
Minor restricted-activity days (age 18–65)	170,000	170,000
Lost work days (age 18–65)	28,000	28,0007
Asthma exacerbation (age 6–18)	15,000	15,000
Hospital admissions—respiratory (all ages)	54	54
Hospital admissions—cardiovascular (age > 18)	66	66
Non-Fatal Heart Attacks (age >18)		
Peters et al. (2001)	230	230
Pooled estimate of 4 studies	25	25

[a] All estimates are rounded to whole numbers with two significant figures. Confidence intervals are unavailable for this analysis because of the benefit-per-ton methodology.

the 12 functions supplied by experts. Figure 7-2 provides a breakdown of monetized benefits by Pollutant. In Table 7-4, we provide the benefits using our anchor points of Krewski et al., and Lepeule et al., as well as the results from the 12 experts' elicitation on PM mortality.

7.2.5   *Characterization of Uncertainty in the Monetized PM$_{2.5}$ Benefits*

In any complex analysis using estimated parameters and inputs from numerous models, there are likely to be many sources of uncertainty. This analysis is no exception. This analysis includes many data sources as inputs, including emission inventories, air quality data from models (with their associated parameters and inputs), population data, population estimates, health effect estimates from epidemiology studies, economic data for monetizing benefits, and assumptions regarding the future state of the world (i.e., regulations, technology, and human behavior). Each of these inputs may be uncertain and would affect the benefits estimate. When the uncertainties from each stage of the analysis are compounded, even small uncertainties can have large effects on the total quantified benefits. Therefore, the estimates of annual benefits

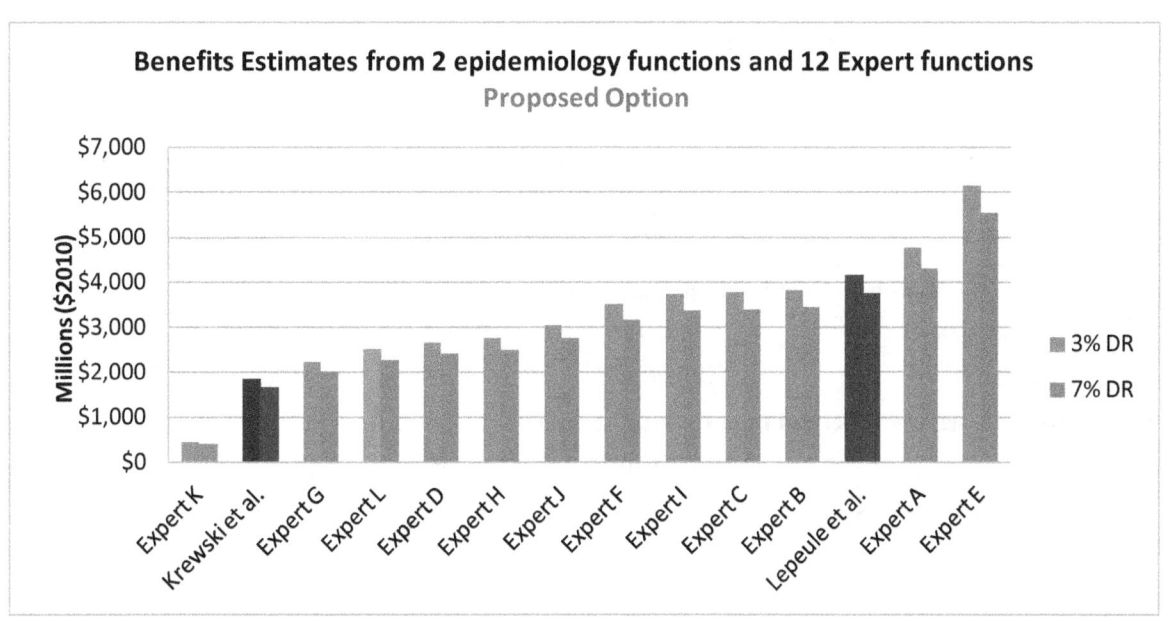

Figure 7-1.   Total Monetized PM$_{2.5}$ Benefits of the Proposed Residential Wood Heaters NSPS in the 2014–2022 Time Frame[a]

[a] This graph shows the estimated benefits at discount rates of 3% and 7% using effect coefficients derived from the Krewski et al. study and the Lepeule et al. study, as well as 12 effect coefficients derived from EPA's expert elicitation on PM mortality. The results shown are not the direct results from the studies or expert elicitation; rather, the estimates are based in part on the concentration-response functions provided in those studies.

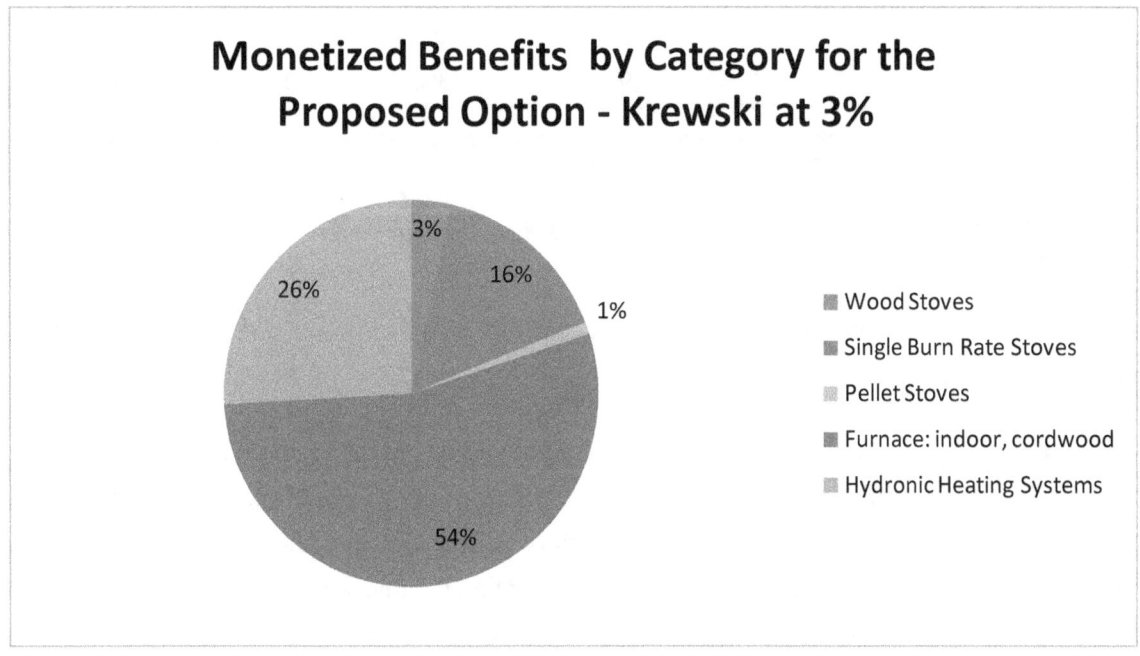

Figure 7-2.   Breakdown of Total Monetized PM$_{2.5}$ Benefits of Proposed Residential Wood Heaters NSPS by Category

Table 7-4. All PM$_{2.5}$ Benefits Estimates for the Proposed Residential Wood Heaters NSPS at Discount Rates of 3% and 7% for the 2014 to 2022 Time Frame ($2010 millions)[a]

	Proposal		Alternative	
	3%	7%	3%	7%
Benefit-per-ton Coefficients Derived from Epidemiology Literature				
Krewski et al.	$1,800	$1,700	$1,900	$1,700
Lepeule et al.	$4,200	$3,700	$4,200	$3,800
Benefit-per-ton Coefficients Derived from Expert Elicitation				
Expert A	$4,800	$4,300	$4,800	$4,300
Expert B	$3,800	$3,400	$3,900	$3,500
Expert C	$3,800	$3,400	$3,800	$3,400
Expert D	$2,700	$2,400	$2,700	$2,500
Expert E	$6,100	$5,500	$6,200	$5,600
Expert F	$3,500	$3,200	$3,600	$3,200
Expert G	$2,200	$2,000	$2,200	$2,000
Expert H	$2,800	$2,500	$2,800	$2,500
Expert I	$3,700	$3,400	$2,800	$3,400
Expert J	$3,000	$2,700	$3,000	$2,800
Expert K	$440	$400	$450	$400
Expert L	$2,500	$2,300	$2,500	$2,300

[a] All estimates are rounded to two significant figures. Estimates do not include confidence intervals because they were derived through the benefit-per-ton technique described above. The benefits estimates from the expert elicitation are provided as a reasonable characterization of the uncertainty in the mortality estimates associated with the concentration-response function. Confidence intervals are unavailable for this analysis because of the benefit-per-ton methodology.

should be viewed as representative of the magnitude of benefits expected, rather than the actual benefits that would occur every year.

This RIA does not include the type of detailed uncertainty assessment found in the PM NAAQS RIA (U.S. EPA, 2012a) because we lack the necessary air quality input and monitoring data to run the benefits model. However, the results of the uncertainty analyses presented in the PM NAAQS RIA can provide some information regarding the uncertainty inherent in the benefits results presented in this analysis. Sensitivity analyses conducted for the PM NAAQS RIA indicate that alternate cessation lag assumptions could change the PM$_{2.5}$-related mortality benefits discounted at 3% by between 10% and −27% and that alternate income growth adjustments could change the $PM_{2.5}$-related mortality benefits by between 33% and −14%.[69]

Unlike the PM NAAQS RIA, we do not have data on the specific location of the air quality changes associated with this rulemaking. As such, it is not feasible to estimate the proportion of benefits occurring in different locations, such as designated nonattainment areas. Instead, we applied benefit-per-ton estimates, which reflect specific geographic patterns of emissions reductions and specific air quality and benefits modeling assumptions. For example, these estimates do not reflect local variability in population density, meteorology, exposure, baseline health incidence rates, or other local factors that might lead to an over-estimate or under-estimate of the actual benefits of controlling PM precursors. Use of these $/ton values to estimate benefits may lead to higher or lower benefit estimates than if benefits were calculated based on direct air quality modeling. Great care should be taken in applying these estimates to emission reductions occurring in any specific location, as these are all based on national or broad regional emission reduction programs and therefore represent average benefits-per-ton over the entire United States. The benefits-per-ton for emission reductions in specific locations may be very different than the estimates presented here. To the extent that the geographic distributions of the emissions reductions for this rule are different than the modeled emissions, the benefits may be underestimated or overestimated. In general, there is inherently more uncertainty for new sources, which may not be included in the emissions inventory, than existing sources. For more information, see the TSD describing the calculation of these benefit-per-ton estimates (U.S. EPA, 2013).

Our estimate of the total benefits is based on EPA's interpretation of the best available scientific literature and methods and supported by the SAB-HES and the NAS (NRC, 2002). Below are key assumptions underlying the estimates for premature mortality, which accounts for 98% of the total monetized $PM_{2.5}$ benefits:

1. We assume that all fine particles, regardless of their chemical composition, are equally potent in causing premature mortality. This is an important assumption, because $PM_{2.5}$ varies considerably in composition across sources, but the scientific evidence is not yet sufficient to allow differentiation of effect estimates by particle type. The PM ISA concluded that "many constituents of $PM_{2.5}$ can be linked with multiple health effects, and the evidence is not yet sufficient to allow differentiation of those constituents or sources that are more closely related to specific outcomes" (U.S. EPA, 2009).

---

[69] http://www.epa.gov/ttn/ecas/regdata/RIAs/finalria.pdf (pp 6-16).

2. We assume that the health impact function for fine particles is log-linear without a threshold in this analysis. Thus, the estimates include health benefits from reducing fine particles in areas with varied concentrations of $PM_{2.5}$, including both areas that do not meet the fine particle standard and those areas that are in attainment, down to the lowest modeled concentrations.

3. We assume that there is a "cessation" lag between the change in PM exposures and the total realization of changes in mortality effects. Specifically, we assume that some of the incidences of premature mortality related to $PM_{2.5}$ exposures occur in a distributed fashion over the 20 years following exposure based on the advice of the SAB-HES (U.S. EPA-SAB, 2004c), which affects the valuation of mortality benefits at different discount rates.

In general, we are more confident in the magnitude of the risks we estimate from simulated $PM_{2.5}$ concentrations that coincide with the bulk of the observed PM concentrations in the epidemiological studies that are used to estimate the benefits. Likewise, we are less confident in the risk we estimate from simulated $PM_{2.5}$ concentrations that fall below the bulk of the observed data in these studies. Concentration benchmark analyses (e.g., lowest measured level [LML] or one standard deviation below the mean of the air quality data in the study) allow readers to determine the portion of population exposed to annual mean $PM_{2.5}$ levels at or above different concentrations, which provides some insight into the level of uncertainty in the estimated $PM_{2.5}$ mortality benefits. There are uncertainties inherent in identifying any particular point at which our confidence in reported associations becomes appreciably less, and the scientific evidence provides no clear dividing line. However, the EPA does not view these concentration benchmarks as a concentration threshold below which we would not quantify health benefits of air quality improvements.[70] Rather, the benefits estimates reported in this RIA are the best estimates because they reflect the full range of air quality concentrations associated with the emission reduction strategies and because the current body of scientific literature indicates that a no-threshold model provides the best estimate of PM-related long-term mortality. In other words, although we may have less confidence in the magnitude of the risk at concentrations below these benchmarks, we still have high confidence that $PM_{2.5}$ is causally associated with risk at those lower air quality concentrations.

For this analysis, policy-specific air quality data is not available due to time or resource limitations. For these rules, we are unable to estimate the percentage of premature mortality associated with this specific rule's emission reductions at each $PM_{2.5}$ level. However, we believe

---

[70] For a summary of the scientific review statements regarding the lack of a threshold in the $PM_{2.5}$-mortality relationship, see the Technical Support Document (TSD) entitled *Summary of Expert Opinions on the Existence of a Threshold in the Concentration-Response Function for PM$_{2.5}$-related Mortality* (U.S. EPA, 2010b).

that it is still important to characterize the distribution of exposure to baseline air quality levels. As a surrogate measure of mortality impacts, we provide the percentage of the population exposed at each $PM_{2.5}$ level in the baseline of the source apportionment modeling used to calculate the benefit-per-ton estimates for this sector. It is important to note that baseline exposure is only one parameter in the health impact function, along with baseline incidence rates population, and change in air quality. In other words, the percentage of the population exposed to air pollution below the LML is not the same as the percentage of the population experiencing health impacts as a result of a specific emission reduction policy. The most important aspect, which we are unable to quantify for rules without rule-specific air quality modeling, is the shift in exposure associated with this specific rule. Therefore, caution is warranted when interpreting the LML assessment for this rule because these results are not consistent with results from rules that had air quality modeling.

Table 7-5 provides the percentage of the population exposed above and below two concentration benchmarks (i.e., LML and 1 standard deviation below the mean) in the modeled baseline. Figure 7-3 shows a bar chart of the percentage of the population exposed to various air quality levels in the baseline, and Figure 7-4 shows a cumulative distribution function of the same data. Both figures identify the LML for each of the major cohort studies.

7.3   Unquantified Benefits

The monetized benefits estimated in this RIA only reflect a subset of benefits attributable to the health effect reductions associated with ambient fine particles. Data, time, and resource limitations prevented EPA from quantifying the impacts to, or monetizing the benefits from several important benefit categories, including benefits associated with the potential exposure to ozone formation due to VOC emissions as a precursor, VOC emissions as a $PM_{2.5}$ precursor,

Table 7-5.   Population Exposure in the Baseline Above and Below Various Concentration Benchmarks in the Underlying Epidemiology Studies[a]

Epidemiology Study	Below 1 Std. Dev. Below AQ Mean	At or Above 1 Std. Dev. Below AQ Mean	Below LML	At or Above LML
Krewski et al. (2009)	89%	11%	7%	93%
Lepeule et al. (2012)	N/A	N/A	23%	67%

[a] One standard deviation below the mean is equivalent to the middle of the range between the 10th and 25th percentile. For Krewski, the LML is 5.8 $\mu g/m^3$ and one standard deviation below the mean is 11.0 $\mu g/m^3$. For Lepeule et al., the LML is 8 $\mu g/m^3$ and we do not have the data for one standard deviation below the mean. It is important to emphasize that although we have lower levels of confidence in levels below the LML for each study, the scientific evidence does not support the existence of a level below which health effects from exposure to $PM_{2.5}$ do not occur.

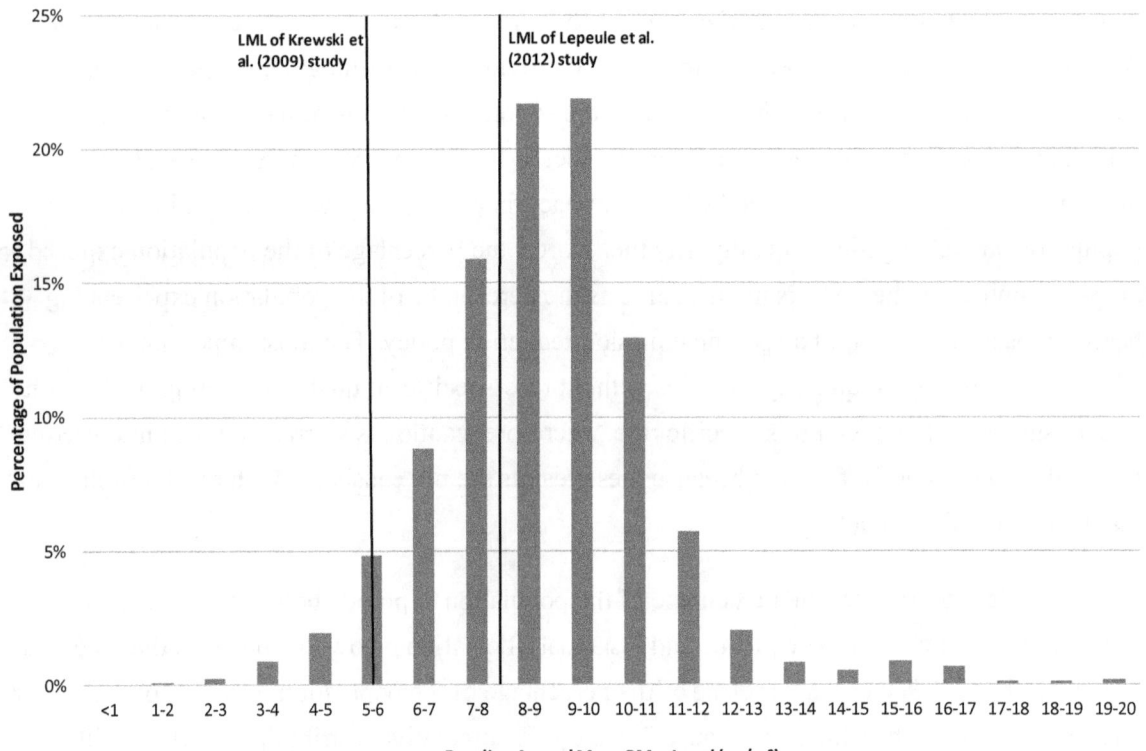

Among the populations exposed to PM$_{2.5}$ in the baseline:
93% are exposed to PM$_{2.5}$ levels at or above the LML of the Krewski et al. (2009) study
67% are exposed to PM$_{2.5}$ levels at or above the LML of the Lepeule et al. (2012) study

Figure 7-3.   Percentage of Adult Population by Annual Mean PM$_{2.5}$ Exposure in the Baseline

HAP, CO exposure, as well as ecosystem effects, and visibility impairment due to the absence of air quality modeling data for these pollutants in this analysis. This does not imply that there are no benefits associated with these emission reductions. In this section, we provide a qualitative description of these benefits.

7.3.1   HAP Benefits

Even though emissions of air toxics from all sources in the U.S. declined by approximately 42% since 1990, the 2005 National-Scale Air Toxics Assessment (NATA) predicts that most Americans are exposed to ambient concentrations of air toxics at levels that have the potential to cause adverse health effects (U.S. EPA, 2011c).[71] The levels of air toxics to

---

[71] The 2005 NATA is available on the Internet at http://www.epa.gov/ttn/atw/nata2005/.

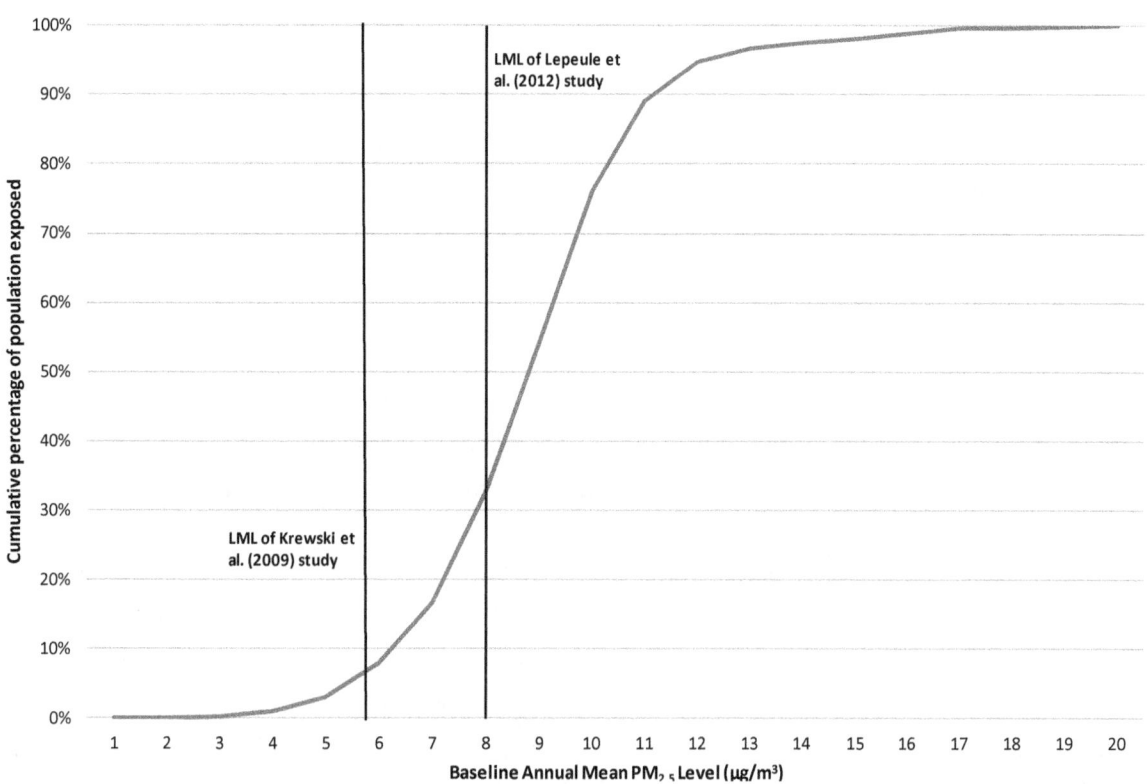

Among the populations exposed to PM$_{2.5}$ in the baseline:
  93% are exposed to PM$_{2.5}$ levels at or above the LML of the Krewski et al. (2009) study
  67% are exposed to PM$_{2.5}$ levels at or above the LML of the Lepeule et al. (2012) study

Figure 7-4.   Cumulative Distribution of Adult Population by Annual Mean PM$_{2.5}$ Exposure in the Baseline

which people are exposed vary depending on where people live and work and the kinds of activities in which they engage. In order to identify and prioritize air toxics, emission source types and locations that are of greatest potential concern, U.S. EPA conducts the NATA.[72] The most recent NATA was conducted for calendar year 2005 and was released in March 2011. NATA includes four steps:

1. Compiling a national emissions inventory of air toxics emissions from outdoor sources

---

[72] The NATA modeling framework has a number of limitations that prevent its use as the sole basis for setting regulatory standards. These limitations and uncertainties are discussed on the 2005 NATA website. Even so, this modeling framework is very useful in identifying air toxic pollutants and sources of greatest concern, setting regulatory priorities, and informing the decision making process. U.S. EPA.(2011). 2005 National-Scale Air Toxics Assessment. http://www.epa.gov/ttn/atw/nata2005/

2. Estimating ambient and exposure concentrations of air toxics across the United States

3. Estimating population exposures across the United States

4. Characterizing potential public health risk due to inhalation of air toxics including both cancer and noncancer effects

Based on the 2005 NATA, EPA estimates that about 5% of census tracts nationwide have increased cancer risks greater than 100 in a million. The average national cancer risk is about 50 in a million. Nationwide, the key pollutants that contribute most to the overall cancer risks are formaldehyde and benzene.[73] Secondary formation (e.g., formaldehyde forming from other emitted pollutants) was the largest contributor to cancer risks, while stationary, mobile and background sources contribute almost equal portions of the remaining cancer risk.

Noncancer health effects can result from chronic,[74] subchronic,[75] or acute[76] inhalation exposures to air toxics, and include neurological, cardiovascular, liver, kidney, and respiratory effects as well as effects on the immune and reproductive systems. According to the 2005 NATA, about three-fourths of the U.S. population was exposed to an average chronic concentration of air toxics that has the potential for adverse noncancer respiratory health effects. Results from the 2005 NATA indicate that acrolein is the primary driver for noncancer respiratory risk.

Figures 7-5 and 7-6 depict the estimated census tract-level carcinogenic risk and noncancer respiratory hazard from the assessment. It is important to note that large reductions in HAP emissions may not necessarily translate into significant reductions in health risk because toxicity varies by pollutant, and exposures may or may not exceed levels of concern. For example, acetaldehyde mass emissions are more than double acrolein emissions on a national basis, according to EPA's 2005 National Emissions Inventory (NEI). However, the Integrated Risk Information System (IRIS) reference concentration (RfC) for acrolein is considerably lower

---

[73] Details about the overall confidence of certainty ranking of the individual pieces of NATA assessments including both quantitative (e.g., model-to-monitor ratios) and qualitative (e.g., quality of data, review of emission inventories) judgments can be found at http://www.epa.gov/ttn/atw/nata/roy/page16.html.

[74] Chronic exposure is defined in the glossary of the Integrated Risk Information (IRIS) database (http://www.epa.gov/iris) as repeated exposure by the oral, dermal, or inhalation route for more than approximately 10% of the life span in humans (more than approximately 90 days to 2 years in typically used laboratory animal species).

[75] Defined in the IRIS database as repeated exposure by the oral, dermal, or inhalation route for more than 30 days, up to approximately 10% of the life span in humans (more than 30 days up to approximately 90 days in typically used laboratory animal species).

[76] Defined in the IRIS database as exposure by the oral, dermal, or inhalation route for 24 hours or less.

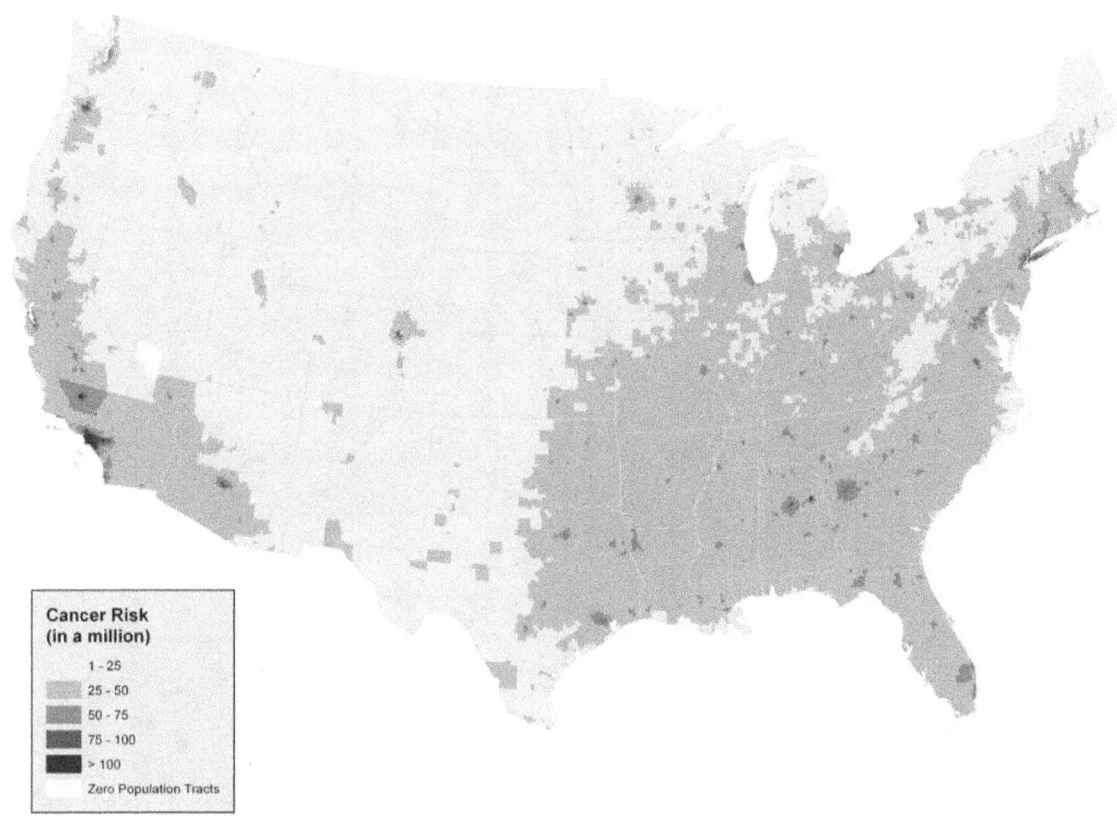

Figure 7-5.   Estimated Census Tract Carcinogenic Risk from HAP Exposure from Outdoor Sources (2005 NATA)

than that for acetaldehyde, suggesting that acrolein could be potentially more toxic than acetaldehyde. Thus, it is important to account for the toxicity and exposure, as well as the mass of the targeted emissions.

Due to methodology limitations, we were unable to estimate the benefits associated with the hazardous air pollutants that would be reduced as a result of these rules. In a few previous analyses of the benefits of reductions in HAP, EPA has quantified the benefits of potential reductions in the incidences of cancer and non-cancer risk (e.g., U.S. EPA, 1995). In those analyses, EPA relied on unit risk factors (URF) developed through risk assessment procedures.[77]These URFs are designed to be conservative, and as such, are more likely to

---

[77] The unit risk factor is a quantitative estimate of the carcinogenic potency of a pollutant, often expressed as the probability of contracting cancer from a 70-year lifetime continuous exposure to a concentration of one $\mu g/m^3$ of a pollutant.

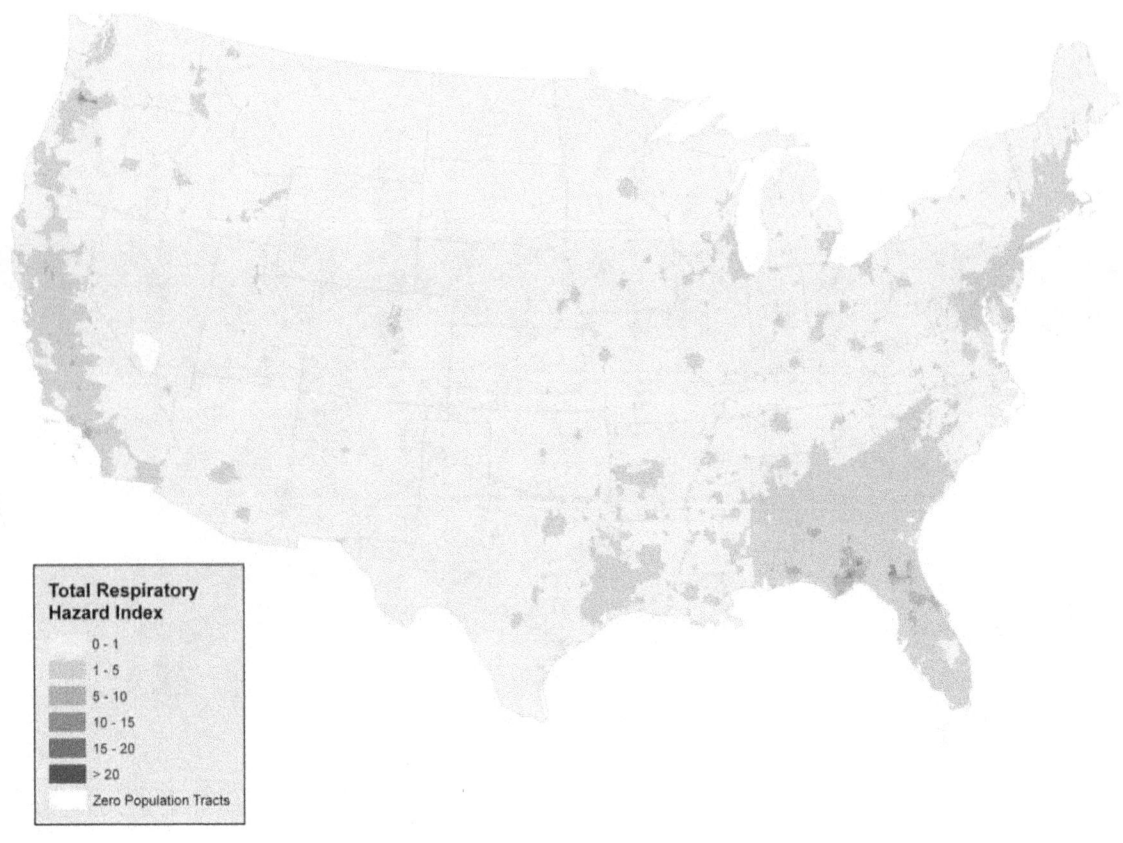

Figure 7-6.   Estimated Chronic Census Tract Noncancer (Respiratory) Risk from HAP Exposure from Outdoor Sources (2005 NATA)

represent the high end of the distribution of risk rather than a best or most likely estimate of risk. As the purpose of a benefit analysis is to describe the benefits most likely to occur from a reduction in pollution, use of high-end, conservative risk estimates would overestimate the benefits of the regulation. While we used high-end risk estimates in past analyses, advice from the EPA's Science Advisory Board (SAB) recommended that we avoid using high-end estimates in benefit analyses (U.S. EPA-SAB, 2002). Since this time, EPA has continued to develop better methods for analyzing the benefits of reductions in HAP.

As part of the second prospective analysis of the benefits and costs of the Clean Air Act (U.S. EPA, 2011a), EPA conducted a case study analysis of the health effects associated with reducing exposure to benzene in Houston from implementation of the Clean Air Act (IEc, 2009). While reviewing the draft report, EPA's Advisory Council on Clean Air Compliance Analysis concluded that "the challenges for assessing progress in health improvement as a result of reductions in emissions of hazardous air pollutants (HAPs) are daunting...due to a lack of exposure-response functions, uncertainties in emissions inventories and background levels, the difficulty of extrapolating risk estimates to low doses and the challenges of tracking health progress for diseases, such as cancer, that have long latency periods" (U.S. EPA-SAB, 2008).

In 2009, EPA convened a workshop to address the inherent complexities, limitations, and uncertainties in current methods to quantify the benefits of reducing HAP. Recommendations from this workshop included identifying research priorities, focusing on susceptible and vulnerable populations, and improving dose-response relationships (Gwinn et al., 2011).

In summary, monetization of the benefits of reductions in cancer incidences requires several important inputs, including central estimates of cancer risks, estimates of exposure to carcinogenic HAP, and estimates of the value of an avoided case of cancer (fatal and non-fatal). Due to methodology limitations, we did not attempt to monetize the health benefits of reductions in HAP in this analysis. Instead, we provide a qualitative analysis of the health effects associated with the HAP anticipated to be reduced by these rules. EPA remains committed to improving methods for estimating HAP benefits by continuing to explore additional concepts of benefits, including changes in the distribution of risk.

Below we describe the health effects associated with the HAPs that would be reduced by this rulemaking.

7.3.1.1 Benzene

The EPA's IRIS database lists benzene as a known human carcinogen (causing leukemia) by all routes of exposure, and concludes that exposure is associated with additional health effects, including genetic changes in both humans and animals and increased proliferation of bone marrow cells in mice.[78,79,80] EPA states in its IRIS database that data indicate a causal relationship between benzene exposure and acute lymphocytic leukemia and suggest a relationship between benzene exposure and chronic non-lymphocytic leukemia and chronic lymphocytic leukemia. The IARC

---

[78] U.S. Environmental Protection Agency (U.S. EPA). 2000. Integrated Risk Information System File for Benzene. Research and Development, National Center for Environmental Assessment, Washington, DC. This material is available electronically at: http://www.epa.gov/iris/subst/0276.htm.

[79] International Agency for Research on Cancer, IARC monographs on the evaluation of carcinogenic risk of chemicals to humans, Volume 29, Some industrial chemicals and dyestuffs, International Agency for Research on Cancer, World Health Organization, Lyon, France, p. 345-389, 1982.

[80] Irons, R.D.; Stillman, W.S.; Colagiovanni, D.B.; Henry, V.A. (1992) Synergistic action of the benzene metabolite hydroquinone on myelopoietic stimulating activity of granulocyte/macrophage colony-stimulating factor in vitro, Proc. Natl. Acad. Sci. 89:3691-3695.

has determined that benzene is a human carcinogen and the DHHS has characterized benzene as a known human carcinogen.[81,82]

A number of adverse noncancer health effects including blood disorders, such as preleukemia and aplastic anemia, have also been associated with long-term exposure to benzene.[83,84]

7.3.1.2 Dioxins (Chlorinated dibenzodioxins (CDDs)[85]

A number of effects have been observed in people exposed to 2,3,7,8-TCDD levels that are at least 10 times higher than background levels. The most obvious health effect in people exposure to relatively large amounts of 2,3,7,8-TCDD is Chloracne. Chloracne is a severe skin disease with acne-like lesions that occur mainly on the face and upper body. Other skin effects noted in people exposed to high doses of 2,3,7,8-TCDD include skin rashes, discoloration, and excessive body hair. Changes in blood and urine that may indicate liver damage also are seen in people. Alterations in the ability of the liver to metabolize (or breakdown) hemoglobin, lipids, sugar, and protein have been reported in people exposed to relatively high concentrations of 2,3,7,8-TCDD. Most of the effects are considered mild and were reversible. However, in some people these effects may last for many years. Slight increases in the risk of diabetes and abnormal glucose tolerance have been observed in some studies of people exposed to 2,3,7,8-TCDD. We do not have enough information to know if exposure to 2,3,7,8-TCDD would result in reproductive or developmental effects in people, but animal studies suggest that this is a potential health concern.

In certain animal species, 2,3,7,8-TCDD is especially harmful and can cause death after a single exposure. Exposure to lower levels can cause a variety of effects in animals, such as weight loss, liver damage, and disruption of the endocrine system. In many species of animals, 2,3,7,8-TCDD weakens the immune system and causes a decrease in the system's ability to fight bacteria and viruses at relatively low levels (approximately 10 times higher than human

---

[81] International Agency for Research on Cancer (IARC). 1987. Monographs on the evaluation of carcinogenic risk of chemicals to humans, Volume 29, Supplement 7, Some industrial chemicals and dyestuffs, World Health Organization, Lyon, France.

[82] U.S. Department of Health and Human Services National Toxicology Program 11th Report on Carcinogens available at: http://ntp.niehs.nih.gov/go/16183 .

[83] Aksoy, M. (1989). Hematotoxicity and carcinogenicity of benzene. Environ. Health Perspect. 82: 193-197.

[84] Goldstein, B.D. (1988). Benzene toxicity. Occupational medicine. State of the Art Reviews. 3: 541-554.

[85] All health effects language for this section came from: Agency for Toxic Substances and Disease Registry (ATSDR). 1999. ToxFAQs for Chlorinated Dibenzo-p-dioxins (CDDs) (CAS#: 2,3,7,8-TCDD 1746-01-6). Atlanta, GA: U.S. Department of Health and Human Services, Public Health Service. Available on the Internet at http://www.atsdr.cdc.gov/tfacts104.html.

background body burdens). In other animal studies, exposure to 2,3,7,8-TCDD has caused reproductive damage and birth defects. Some animal species exposed to CDDs during pregnancy had miscarriages and the offspring of animals exposed to 2,3,7,8-TCDD during pregnancy often had severe birth defects including skeletal deformities, kidney defects, and weakened immune responses. In some studies, effects were observed at body burdens 10 times higher than human background levels.

7.3.1.3 Formaldehyde

Since 1987, EPA has classified formaldehyde as a probable human carcinogen based on evidence in humans and in rats, mice, hamsters, and monkeys.[86] Substantial additional research since that time informs current scientific understanding of the health effects associated with exposure to formaldehyde. These include recently published research conducted by the National Cancer Institute (NCI) which found an increased risk of nasopharyngeal cancer and lymphohematopoietic malignancies such as leukemia among workers exposed to formaldehyde.[87,88] In an analysis of the lymphohematopoietic cancer mortality from an extended follow-up of these workers, NCI confirmed an association between lymphohematopoietic cancer risk and peak formaldehyde exposures.[89] A recent NIOSH study of garment workers also found increased risk of death due to leukemia among workers exposed to formaldehyde.[90] Extended follow-up of a cohort of British chemical workers did not find evidence of an increase in nasopharyngeal or lymphohematopoietic cancers, but a continuing statistically significant excess in lung cancers was reported.[91]

In the past 15 years there has been substantial research on the inhalation dosimetry for formaldehyde in rodents and primates by the Chemical Industry Institute of Toxicology (CIIT, now renamed the Hamner Institutes for Health Sciences), with a focus on use of rodent data for

---

[86] U.S. EPA. 1987. Assessment of Health Risks to Garment Workers and Certain Home Residents from Exposure to Formaldehyde, Office of Pesticides and Toxic Substances, April 1987. Docket EPA-HQ-OAR-2010-0162.

[87] Hauptmann, M.; Lubin, J. H.; Stewart, P. A.; Hayes, R. B.; Blair, A. 2003. Mortality from lymphohematopoetic malignancies among workers in formaldehyde industries. Journal of the National Cancer Institute 95: 1615-1623. Docket EPA-HQ-OAR-2010-0162.

[88] Hauptmann, M.; Lubin, J. H.; Stewart, P. A.; Hayes, R. B.; Blair, A. 2004. Mortality from solid cancers among workers in formaldehyde industries. American Journal of Epidemiology 159: 1117-1130. Docket EPA-HQ-OAR-2010-0162.

[89] Beane Freeman, L. E.; Blair, A.; Lubin, J. H.; Stewart, P. A.; Hayes, R. B.; Hoover, R. N.; Hauptmann, M. 2009. Mortality from lymphohematopoietic malignancies among workers in formaldehyde industries: The National Cancer Institute cohort. J. National Cancer Inst. 101: 751-761. Docket EPA-HQ-OAR-2010-0162.

[90] Pinkerton, L. E. 2004. Mortality among a cohort of garment workers exposed to formaldehyde: an update. Occup. Environ. Med. 61: 193-200. Docket EPA-HQ-OAR-2010-0162.

[91] Coggon, D, EC Harris, J Poole, KT Palmer. 2003. Extended follow-up of a cohort of British chemical workers exposed to formaldehyde. J National Cancer Inst. 95:1608-1615. Docket EPA-HQ-OAR-2010-0162.

refinement of the quantitative cancer dose-response assessment.[92,93,94] CIIT's risk assessment of formaldehyde incorporated mechanistic and dosimetric information on formaldehyde. These data were modeled using a biologically-motivated two-stage clonal growth model for cancer and also a point of departure based on a Benchmark Dose approach. However, it should be noted that recent research published by EPA indicates that when two-stage modeling assumptions are varied, resulting dose-response estimates can vary by several orders of magnitude.[95,96,97,98] These findings are not supportive of interpreting the CIIT model results as providing a conservative (health protective) estimate of human risk.[99] EPA research also examined the contribution of the two-stage modeling for formaldehyde towards characterizing the relative weights of key events in the mode-of-action of a carcinogen. For example, the model-based inference in the published CIIT study that formaldehyde's direct mutagenic action is not relevant to the compound's tumorigenicity was found not to hold under variations of modeling assumptions.[100]

Based on the developments of the last decade, in 2004, the working group of the IARC concluded that formaldehyde is carcinogenic to humans (Group 1), on the basis of sufficient evidence in humans and sufficient evidence in experimental animals—a higher classification than previous IARC evaluations. After reviewing the currently available epidemiological evidence, the IARC (2006) characterized the human evidence for formaldehyde carcinogenicity

---

[92] Conolly, RB, JS Kimbell, D Janszen, PM Schlosser, D Kalisak, J Preston, and FJ Miller. 2003. Biologically motivated computational modeling of formaldehyde carcinogenicity in the F344 rat. Tox Sci 75: 432-447. Docket EPA-HQ-OAR-2010-0162.

[93] Conolly, RB, JS Kimbell, D Janszen, PM Schlosser, D Kalisak, J Preston, and FJ Miller. 2004. Human respiratory tract cancer risks of inhaled formaldehyde: Dose-response predictions derived from biologically-motivated computational modeling of a combined rodent and human dataset. Tox Sci 82: 279-296. Docket EPA-HQ-OAR-2010-0162.

[94] Chemical Industry Institute of Toxicology (CIIT).1999. Formaldehyde: Hazard characterization and dose-response assessment for carcinogenicity by the route of inhalation. CIIT, September 28, 1999. Research Triangle Park, NC. Docket EPA-HQ-OAR-2010-0162.

[95] U.S. EPA. Analysis of the Sensitivity and Uncertainty in 2-Stage Clonal Growth Models for Formaldehyde with Relevance to Other Biologically-Based Dose Response (BBDR) Models. U.S. Environmental Protection Agency, Washington, D.C., EPA/600/R-08/103, 2008. Docket EPA-HQ-OAR-2010-0162.

[96] Subramaniam, R; Chen, C; Crump, K; et al. (2008) Uncertainties in biologically-based modeling of formaldehyde-induced cancer risk: identification of key issues. Risk Anal 28(4):907-923. Docket EPA-HQ-OAR-2010-0162.

[97] Subramaniam RP; Crump KS; Van Landingham C; et al. (2007) Uncertainties in the CIIT model for formaldehyde-induced carcinogenicity in the rat: A limited sensitivity analysis–I. Risk Anal, 27: 1237–1254. Docket EPA-HQ-OAR-2010-0162.

[98] Crump, K; Chen, C; Fox, J; et al. (2008) Sensitivity analysis of biologically motivated model for formaldehyde-induced respiratory cancer in humans. Ann Occup Hyg 52:481-495. Docket EPA-HQ-OAR-2010-0162.

[99] Crump, K; Chen, C; Fox, J; et al. (2008) Sensitivity analysis of biologically motivated model for formaldehyde-induced respiratory cancer in humans. Ann Occup Hyg 52:481-495. Docket EPA-HQ-OAR-2010-0162.

[100] Subramaniam RP; Crump KS; Van Landingham C; et al. (2007) Uncertainties in the CIIT model for formaldehyde-induced carcinogenicity in the rat: A limited sensitivity analysis–I. Risk Anal, 27: 1237–1254. Docket EPA-HQ-OAR-2010-0162.

as "sufficient," based upon the data on nasopharyngeal cancers; the epidemiologic evidence on leukemia was characterized as "strong."[101]

Formaldehyde exposure also causes a range of noncancer health effects, including irritation of the eyes (burning and watering of the eyes), nose and throat. Effects from repeated exposure in humans include respiratory tract irritation, chronic bronchitis and nasal epithelial lesions such as metaplasia and loss of cilia. Animal studies suggest that formaldehyde may also cause airway inflammation—including eosinophils infiltration into the airways. There are several studies that suggest that formaldehyde may increase the risk of asthma—particularly in the young.[102,103]

The above-mentioned rodent and human studies, as well as mechanistic information and their analyses, were evaluated in EPA's recent Draft Toxicological Review of Formaldehyde—Inhalation Assessment through the Integrated Risk Information System (IRIS) program. This draft IRIS assessment was released in June 2010 for public review and comment and external peer review by the National Research Council (NRC). The NRC released their review report in April 2011 (http://www.nap.edu/catalog.php?record_id=13142). The EPA is currently revising the draft assessment in response to this review.

7.3.1.4 Polycyclic Organic Matter (POM)

The term polycyclic organic matter (POM) defines a broad class of compounds that includes the polycyclic aromatic hydrocarbon compounds (PAHs). One of these compounds, naphthalene, is discussed separately below. POM compounds are formed primarily from combustion and are present in the atmosphere in gas and particulate form. Cancer is the major concern from exposure to POM. Epidemiologic studies have reported an increase in lung cancer in humans exposed to diesel exhaust, coke oven emissions, roofing tar emissions, and cigarette smoke; all of these mixtures contain POM compounds[104,105.] Animal studies have reported

---

[101] International Agency for Research on Cancer (2006) Formaldehyde, 2-Butoxyethanol and 1-tert-Butoxypropan-2-ol. Monographs Volume 88. World Health Organization, Lyon, France. Docket EPA-HQ-OAR-2010-0162.

[102] Agency for Toxic Substances and Disease Registry (ATSDR). 1999. Toxicological profile for Formaldehyde. Atlanta, GA: U.S. Department of Health and Human Services, Public Health Service. http://www.atsdr.cdc.gov/toxprofiles/tp111.html. Docket EPA-HQ-OAR-2010-0162.

[103] WHO (2002) Concise International Chemical Assessment Document 40: Formaldehyde. Published under the joint sponsorship of the United Nations Environment Programme, the International Labour Organization, and the World Health Organization, and produced within the framework of the Inter-Organization Programme for the Sound Management of Chemicals. Geneva. Docket EPA-HQ-OAR-2010-0162.

[104] Agency for Toxic Substances and Disease Registry (ATSDR). 1995. Toxicological profile for Polycyclic Aromatic Hydrocarbons (PAHs). Atlanta, GA: U.S. Department of Health and Human Services, Public Health Service. Available electronically at http://www.atsdr.cdc.gov/ToxProfiles/TP.asp?id=122&tid=25.

respiratory tract tumors from inhalation exposure to benzo[a]pyrene and alimentary tract and liver tumors from oral exposure to benzo[a]pyrene. EPA has classified seven PAHs (benzo[a]pyrene, benz[a]anthracene, chrysene, benzo[b]fluoranthene, benzo[k]fluoranthene, dibenz[a,h]anthracene, and indeno[1,2,3-cd]pyrene) as Group B2, probable human carcinogens.[106] Recent studies have found that maternal exposures to PAHs in a population of pregnant women were associated with several adverse birth outcomes, including low birth weight and reduced length at birth, as well as impaired cognitive development in preschool children (3 years of age).[107,108] EPA has not yet evaluated these recent studies.

7.3.1.5 Other Air Toxics

In addition to the compounds described above, other compounds would be affected by this rule. Information regarding the health effects of these compounds can be found in EPA's IRIS database.[109]

7.3.2 Carbon Monoxide Co-Benefits

Carbon monoxide in ambient air is formed primarily by the incomplete combustion of carbon-containing fuels and photochemical reactions in the atmosphere. The amount of CO emitted from these reactions, relative to carbon dioxide (CO_2), is sensitive to conditions in the combustion zone, such as fuel oxygen content, burn temperature, or mixing time. Upon inhalation, CO diffuses through the respiratory system to the blood, which can cause hypoxia (reduced oxygen availability). Carbon monoxide can elicit a broad range of effects in multiple tissues and organ systems that are dependent upon concentration and duration of exposure. The *Integrated Science Assessment for Carbon Monoxide* (U.S. EPA, 2010a) concluded that short-term exposure to CO is "likely to have a causal relationship" with cardiovascular morbidity, particularly in individuals with coronary heart disease. Epidemiologic studies associate short-term CO exposure with increased risk of emergency department visits and hospital admissions. Coronary heart disease includes those who have angina pectoris (cardiac chest pain), as well as

[105] U.S. EPA (2002). Health Assessment Document for Diesel Engine Exhaust. EPA/600/8-90/057F Office of Research and Development, Washington DC. Retrieved on March 17, 2009 from http://cfpub.epa.gov/ncea/cfm/recordisplay.cfm?deid=29060. Docket EPA-HQ-OAR-2010-0162.

[106] U.S. EPA (1997). Integrated Risk Information System File of indeno(1,2,3-cd)pyrene. Research and Development, National Center for Environmental Assessment, Washington, DC. This material is available electronically at http://www.epa.gov/ncea/iris/subst/0457.htm.

[107] Perera, F.P.; Rauh, V.; Tsai, W-Y.; et al. (2002) Effect of transplacental exposure to environmental pollutants on birth outcomes in a multiethnic population. Environ Health Perspect. 111: 201-205.

[108] Perera, F.P.; Rauh, V.; Whyatt, R.M.; Tsai, W.Y.; Tang, D.; Diaz, D.; Hoepner, L.; Barr, D.; Tu, Y.H.; Camann, D.; Kinney, P. (2006) Effect of prenatal exposure to airborne polycyclic aromatic hydrocarbons on neurodevelopment in the first 3 years of life among inner-city children. Environ Health Perspect 114: 1287-1292.

[109] U.S. EPA Integrated Risk Information System (IRIS) database is available at: www.epa.gov/iris.

those who have experienced a heart attack. Other subpopulations potentially at risk include individuals with diseases such as chronic obstructive pulmonary disease (COPD), anemia, or diabetes, and individuals in very early or late life stages, such as older adults or the developing young. The evidence is suggestive of a causal relationship between short-term exposure to CO and respiratory morbidity and mortality. The evidence is also suggestive of a causal relationship for birth outcomes and developmental effects following long-term exposure to CO, and for central nervous system effects linked to short- and long-term exposure to CO.

7.3.3 Black Carbon (BC) Benefits

Incomplete combustion of wood results in emissions of fine and ultrafine particles, including black carbon (BC), brown carbon (BrC), and other nonlight, absorbing organic carbon (OC) particles. BC and BrC are collectively considered light, absorbing carbon (LAC) with BC referring to the most strongly light-absorbing form of carbon per unit mass. BC impacts the earth's climate because of its high capacity for light absorption. The role of BC in key atmospheric processes links it to a range of climate impacts, including increased temperatures, accelerated ice and snow melt, and disruptions in precipitation patterns. A recent study by the UN Environment Programme (UNEP) concluded that reductions in BC and ozone will slow the rate of climate change within the first half of this century with a small number of targeted BC and ozone precursor emissions mitigation measures providing immediate protection for climate, public health, water and food security, and ecosystems (UNEP, 2011).[110]

While less effective in absorbing solar radiation than BC, BrC may contribute significantly to positive radiative forcing. At present the ability to quantify the climate impacts of BrC is limited. OC from incomplete combustion of wood (exclusive of BrC) is generally considered nonlight-absorbing carbon. Nonlight absorbing compounds scatter rather than absorb solar radiation and, therefore, provide a net direct cooling effect on climate. Thus, particles generated by residential wood combustion consist of components that are warming to the atmosphere (BC and BrC) and particles that are cooling (OC exclusive of BrC).

Residential wood combustion contributed about 380,000 tons of $PM_{2.5}$ emissions across the United States in 2005. Of these $PM_{2.5}$ emissions, approximately 21,000 tons are estimated to

---

[110] UN Environment Programme, World Meteorological Organization. 2011, February. *Integrated Assessment of Black Carbon and Tropospheric Ozone: Summary for Decision Makers.* Available at http://www.unep.org/gc/gc26/download.asp?ID=2197.

be elemental carbon (EC)[111] (EPA NEI, 2005).[112]

The EC/OC ratio is a metric sometimes used to crudely compare the warming potential of emissions from various BC sources with a ratio of less than 1 indicating that cooling potential exceeds warming. Based on the speciated 2005 NEI, the EC/OC for residential wood combustion is estimated to be less than one (~ 0.11), indicating a predominance of OC or light-scattering particles relative to light absorbing ones. Exactly how much of the OC from RWC sources is light absorbing (BrC) is not known currently, and the LAC may vary by fuel type, combustion conditions, and operating environment.

While OC emissions are generally considered to have a cooling effect, OC emissions over areas with snow and ice may be less reflective than OC over dark surfaces and may even have a slight warming effect (Flanner et al., 2007).[113] Significantly, the vast majority of residential wood smoke emissions occur during the winter months; the highest percentage of wood stove use is in the upper Midwest (e.g., Michigan), the Northeast (e.g., Maine), and the mountainous areas of the Pacific Northwest (e.g., Washington), where snow is present a good portion of the winter months. A recent study of the effect of soot-induced snow albedo on snowpack and hydrological cycles in the western United States concludes that radiative forcing induced by soot deposition on snow is an important anthropogenic source affecting the global climate. The study concludes that soot-induced snow albedo perturbations increase the surface net solar radiation flux during late winter to early spring, increase the surface air temperature, reduce the snow accumulation and spring snowmelt, and may alter stream flows with implications for water resources in the western United States (Qian, et al., 2009).[114] Further study is needed to better understand and quantify the impact of PM$_{2.5}$ emissions and deposition from the RWC sector on climate.

---

[111] BC is roughly equivalent to 'soot carbon' or the portion of soot that is closest to elemental carbon. The most commonly used measurement technique, the 'thermal optical method' quantifies the portion of PM that is EC. EC is frequently used for emissions characterization and ambient measurements. The terms EC and BC are used interchangeably in this discussion.

[112] U.S. EPA. 2005. *National Emissions Inventory.* 2005 Modeling Inventory. Available at. http://www.epa.gov/ttn/chief/emch/index.html.

[113] Flanner, M. G., Zender, C. S., Randerson, J. T., and Rasch, P. J. 2007. Present-day climate forcing and response from BC in snow. *Journal of Geophysical Research-Atmospheres,* 12(D11). doi:10.1029/2006JD008003

[114] Qian, Y., W. I. Gustafson, L. R. Leung, and S. J. Ghan. 2009. Effects of soot-induced snow albedo change on snowpack and hydrological cycle in western United States based on Weather Research and Forecasting chemistry and regional climate simulations, *J. Geophys. Res.* 114, D03108. doi:10.1029/2008JD011039

7.3.4   VOCs as a PM$_{2.5}$ Precursor

This rulemaking is expected to reduce emissions of VOCs, which are a precursor to PM$_{2.5}$. Most VOCs emitted are oxidized to carbon dioxide (CO_2) rather than to PM, but a portion of VOC emission contributes to ambient PM$_{2.5}$ levels as organic carbon aerosols (U.S. EPA, 2009c). Therefore, reducing these emissions would reduce PM$_{2.5}$ formation, human exposure to PM$_{2.5}$, and the incidence of PM$_{2.5}$-related health effects. However, we have not quantified the PM$_{2.5}$-related benefits associated with VOC reductions in this analysis. Analysis of organic carbon measurements suggest only a fraction of secondarily formed organic carbon aerosols are of anthropogenic origin. The current state of the science of secondary organic carbon aerosol formation indicates that anthropogenic VOC contribution to secondary organic carbon aerosol is often lower than the biogenic (natural) contribution. Given that a fraction of secondarily formed organic carbon aerosols is from anthropogenic VOC emissions and the extremely small amount of VOC emissions from this sector relative to the entire VOC inventory it is unlikely this sector has a large contribution to ambient secondary organic carbon aerosols. Photochemical models typically estimate secondary organic carbon from anthropogenic VOC emissions to be less than 0.1 $\mu g/m^3$.

Due to limited resources, we were unable to perform air quality modeling for this rule. Therefore, given the high degree of variability in the responsiveness of PM$_{2.5}$ formation to VOC emission reductions, we are unable to estimate the effect that reducing VOCs will have on ambient PM$_{2.5}$ levels without air quality modeling.

7.3.5   VOCs as an Ozone Precursor

In the presence of sunlight, VOCs can undergo a chemical reaction in the atmosphere to form ozone. Reducing ambient ozone concentrations is associated with significant human health benefits, including mortality and respiratory morbidity (U.S. EPA, 2008a). Epidemiological researchers have associated ozone exposure with adverse health effects in numerous toxicological, clinical and epidemiological studies (U.S. EPA, 2006). These health effects include respiratory morbidity such as fewer asthma attacks, hospital and ER visits, school loss days, as well as premature mortality.

In addition to health impacts reduction, there are ecological benefits from reducing the formation of ozone and related exposure that leads to reduced net primary productivity and visible foliar injury which are associated with a range of ecosystems services.

7.3.6   Visibility Impairment Co-Benefits

Reducing secondary formation of $PM_{2.5}$ would improve visibility levels in the U.S. because suspended particles and gases degrade visibility by scattering and absorbing light (U.S. EPA, 2009). Fine particles with significant light-extinction efficiencies include sulfates, nitrates, organic carbon, elemental carbon, and soil (Sisler, 1996). Visibility has direct significance to people's enjoyment of daily activities and their overall sense of wellbeing. Good visibility increases the quality of life where individuals live and work, and where they engage in recreational activities. Particulate sulfate is the dominant source of regional haze in the eastern U.S. and particulate nitrate is an important contributor to light extinction in California and the upper Midwestern U.S., particularly during winter (U.S. EPA, 2009). Previous analyses (U.S. EPA, 2011a) show that visibility benefits can be a significant welfare benefit category. Without air quality modeling, we are not unable to estimate visibility related benefits, nor are we able to determine whether the emission reductions associated with this rule would be likely to have a significant impact on visibility in urban areas or Class I areas.

7.4   References

Abt Associates, Inc. 2012. *BenMAP User's Manual Appendices*. Prepared for U.S. Environmental Protection Agency Office of Air Quality Planning and Standards. Research Triangle Park, NC. September. Available on the Internet at http://www.epa.gov/air/benmap/models/BenMAPAppendicesOct2012.pdf.

Fann, N., C.M. Fulcher, B.J. Hubbell. 2009. "The influence of location, source, and emission type in estimates of the human health benefits of reducing a ton of air pollution." *Air Qual Atmos Health* 2:169–176.

Fann, N., K.R. Baker, and C.M. Fulcher. 2012. "Characterizing the $PM_{2.5}$-related health benefits of emission reductions for 17 industrial, area and mobile emission sectors across the U.S." *Environment International* 49 41–151.

Gwinn, M.R., J. Craig, D.A. Axelrad, R. Cook, C. Dockins, N. Fann, R. Fegley, D.E. Guinnup, G. Helfand, B. Hubbell, S.L. Mazur, T. Palma, R.L. Smith, J. Vandenberg, and B. Sonawane. 2011. "Meeting report: Estimating the benefits of reducing hazardous air pollutants—summary of 2009 workshop and future considerations." *Environ Health Perspect.* Jan; 119(1): p. 125-30.

Industrial Economics, Inc (IEc). 2009. *Section 812 Prospective Study of the Benefits and Costs of the Clean Air Act: Air Toxics Case Study—Health Benefits of Benzene Reductions in Houston, 1990–2020*. Final Report, July 14, 2009. Available on the Internet at http://www.epa.gov/air/sect812/dec09/812CAAA_Benzene_Houston_Final_Report_July_2009.pdf.

Industrial Economics, Incorporated (IEc). 2006. *Expanded Expert Judgment Assessment of the Concentration-Response Relationship Between PM$_{2.5}$ Exposure and Mortality*. Prepared for: Office of Air Quality Planning and Standards, U.S. Environmental Protection Agency, Research Triangle Park, NC. September. Available on the Internet at http://www.epa.gov/ttn/ecas/regdata/Uncertainty/pm_ee_tsd_expert_interview_summaries.pdf.

Krewski D, Jerrett M, Burnett RT, Ma R, Hughes E, Shi, Y, et al. 2009. *Extended follow-up and spatial analysis of the American Cancer Society study linking particulate air pollution and mortality*. HEI Research Report, 140, Health Effects Institute, Boston, MA.

Lepeule J, Laden F, Dockery D, Schwartz J 2012. "Chronic Exposure to Fine Particles and Mortality: An Extended Follow-Up of the Harvard Six Cities Study from 1974 to 2009." *Environ Health Perspect.* Jul;120(7):965-70.

National Research Council (NRC). 2002. *Estimating the Public Health Benefits of Proposed Air Pollution Regulations*. Washington, DC: The National Academies Press.

Office of Management and Budget (OMB). 2003. *Circular A-4: Regulatory Analysis*. Washington, DC. Available on the Internet at http://www.whitehouse.gov/omb/circulars/a004/a-4.html.

Roman, Henry A., Katherine D. Walker, Tyra L. Walsh, Lisa Conner, Harvey M. Richmond, Bryan J. Hubbell, and Patrick L. Kinney. 2008. "Expert Judgment Assessment of the Mortality Impact of Changes in Ambient Fine Particulate Matter in the U.S." *Environ. Sci. Technol.*, 42(7):2268-2274.

Sisler, J.F. 1996. *Spatial and seasonal patterns and long-term variability of the composition of the haze in the United States: an analysis of data from the IMPROVE network*. CIRA Report, ISSN 0737-5352-32, Colorado State University.

U.S. Census Bureau. 2008a. Firm Size Data from the Statistics of U.S. Businesses: U.S. Detail Employment Sizes: 2002. <http://www2.census.gov/csd/susb/2002/02us_detailed%20sizes_6digitnaics.txt >

U.S. Environmental Protection Agency—Science Advisory Board (U.S. EPA-SAB). 2002. *Workshop on the Benefits of Reductions in Exposure to Hazardous Air Pollutants: Developing Best Estimates of Dose-Response Functions An SAB Workshop Report of an EPA/SAB Workshop (Final Report)*. EPA-SAB-EC-WKSHP-02-001. January. Available on the Internet at http://yosemite.epa.gov/sab%5CSABPRODUCT.NSF/34355712EC011A358525719A005BF6F6/$File/ecwkshp02001%2Bappa-g.pdf.

U.S. Environmental Protection Agency (U.S. EPA). 1995. *Regulatory Impact Analysis for the Petroleum Refinery NESHAP. Revised Draft for Promulgation*. Office of Air Quality Planning and Standards, Research Triangle Park, N.C. Available on the Internet at http://www.reg-markets.org/admin/authorpdfs/page.php?id=705.

U.S. Environmental Protection Agency (U.S. EPA). 2006. *Air Quality Criteria for Ozone and Related Photochemical Oxidants (Final)*. EPA/600/R-05/004aF-cF. Washington, DC: U.S. EPA. February. Available on the Internet at http://cfpub.epa.gov/ncea/CFM/recordisplay.cfm?deid=149923.

U.S. Environmental Protection Agency (U.S. EPA). 2008a. *Integrated Science Assessment for Sulfur Oxides—Health Criteria (Final Report)*. National Center for Environmental Assessment, Research Triangle Park, NC. September. Available on the Internet at http://cfpub.epa.gov/ncea/cfm/recordisplay.cfm?deid=198843.

U.S. Environmental Protection Agency (U.S. EPA). 2008c. *Integrated Science Assessment for Oxides of Nitrogen—Health Criteria (Final Report)*. National Center for Environmental Assessment, Research Triangle Park, NC. July. Available at at http://cfpub.epa.gov/ncea/cfm/recordisplay.cfm?deid=194645.

U.S. Environmental Protection Agency (U.S. EPA). 2008d. *Integrated Science Assessment for Oxides of Nitrogen and Sulfur–Ecological Criteria National (Final Report)*. National Center for Environmental Assessment, Research Triangle Park, NC. EPA/600/R-08/139. December. Available at http://cfpub.epa.gov/ncea/cfm/recordisplay.cfm?deid=201485.

U.S. Environmental Protection Agency (U.S. EPA). 2009. *Integrated Science Assessment for Particulate Matter (Final Report)*. EPA-600-R-08-139F. National Center for Environmental Assessment—RTP Division. December. Available on the Internet at http://cfpub.epa.gov/ncea/cfm/recordisplay.cfm?deid=216546.

U.S. Environmental Protection Agency (U.S. EPA). 2010a. *Integrated Science Assessment for Carbon Monoxide*. National Center for Environmental Assessment, Research Triangle Park, NC. EPA/600/R-09/019F. January. Available at http://cfpub.epa.gov/ncea/cfm/recordisplay.cfm?deid=218686.

U.S. Environmental Protection Agency (U.S. EPA). 2010b. *Technical Support Document: Summary of Expert Opinions on the Existence of a Threshold in the Concentration-Response Function for $PM_{2.5}$-related Mortality*. Research Triangle Park, NC. June. Available on the Internet at www.epa.gov/ttn/ecas/regdata/Benefits/thresholdstsd.pdf.

U.S. Environmental Protection Agency (U.S. EPA). 2010c. *Valuing Mortality Risk Reductions for Environmental Policy: A White Paper: SAB Review Draft*. National Center for Environmental Economics December. Available on the Internet at http://yosemite.epa.gov/ee/epa/eerm.nsf/vwAN/EE-0563-1.pdf/$file/EE-0563-1.pdf.

U.S. Environmental Protection Agency (U.S. EPA). 2010e. *Guidelines for Preparing Economic Analyses*. EPA 240-R-10-001. National Center for Environmental Economics, Office of Policy Economics and Innovation. Washington, DC. December. Available on the Internet at http://yosemite.epa.gov/ee/epa/eerm.nsf/vwAN/EE-0568-50.pdf/$file/EE-0568-50.pdf.

U.S. Environmental Protection Agency (U.S. EPA). 2011a. *The Benefits and Costs of the Clean Air Act from 1990 to 2020*. Office of Air and Radiation, Washington, DC. March. http://www.epa.gov/air/sect812/feb11/fullreport.pdf. Accessed March 30, 2011.

U.S. Environmental Protection Agency (U.S. EPA). 2011b. *Regulatory Impact Analysis for the Final Mercury and Air Toxics Standards.* EPA-452/R-11-011. December. Available on the Internet at http://www.epa.gov/ttn/ecas/regdata/RIAs/matsriafinal.pdf.

U.S. Environmental Protection Agency (U.S. EPA). 2011c. *2005 National-Scale Air Toxics Assessment.* Office of Air Quality Planning and Standards, Research Triangle Park, NC. March. Available on the Internet at http://www.epa.gov/ttn/atw/nata2005/.

U.S. Environmental Protection Agency (U.S. EPA). 2011d. *Regulatory Impact Analysis: National Emission Standards for Hazardous Air Pollutants for Industrial, Commercial, and Institutional Boilers and Process Heaters.* February. Available on the Internet at http://www.epa.gov/ttnecas1/regdata/RIAs/boilersriafinal110221_psg.pdf.

U.S. Environmental Protection Agency (U.S. EPA). 2012a. *Regulatory Impact Analysis for the Final Revisions to the National Ambient Air Quality Standards for Particulate Matter.* EPA-452/R-12-003. Office of Air Quality Planning and Standards, Health and Environmental Impacts Division. December. Available at http://www.epa.gov/pm/2012/finalria.pdf.

U.S. Environmental Protection Agency (U.S. EPA). 2012b. *Regulatory Impact Analysis: Petroleum Refineries New Source Performance Standards Ja.* Office of Air Quality Planning and Standards, Health and Environmental Impacts Division. June. Available at http://www.epa.gov/ttnecas1/regdata/RIAs/refineries_nsps_ja_final_ria.pdf.

U.S. Environmental Protection Agency (U.S. EPA). 2013. *Technical Support Document: Estimating the Benefit per ton of Reducing $PM_{2.5}$ Precursors from 17 sectors.* Office of Air Quality Planning and Standards, Research Triangle Park, NC. February.

U.S. Environmental Protection Agency—Science Advisory Board (U.S. EPA-SAB). 2008. *Characterizing Uncertainty in Particulate Matter Benefits Using Expert Elicitation.* EPA-COUNCIL-08-002. July. Available on the Internet at http://yosemite.epa.gov/sab/sabproduct.nsf/0/43B91173651AED9E85257487004EA6CB/ $File/EPA-COUNCIL-08-002-unsigned.pdf.

U.S. Environmental Protection Agency—Science Advisory Board (U.S. EPA-SAB). 2000. *An SAB Report on EPA's White Paper Valuing the Benefits of Fatal Cancer Risk Reduction.* EPA-SAB-EEAC-00-013. July. Available on the Internet at http://yosemite.epa.gov/sab%5CSABPRODUCT.NSF/41334524148BCCD6852571A700 516498/$File/eeacf013.pdf.

U.S. Environmental Protection Agency—Science Advisory Board (U.S. EPA-SAB). 2004c. *Advisory Council on Clean Air Compliance Analysis Response to Agency Request on Cessation Lag.* EPA-COUNCIL-LTR-05-001. December. Available on the Internet at <. http://yosemite.epa.gov/sab/sabproduct.nsf/0/39F44B098DB49F3C85257170005293E0/$ File/council_ltr_05_001.pdf

U.S. Environmental Protection Agency—Science Advisory Board (U.S. EPA-SAB). 2008.
Benefits of Reducing Benzene Emissions in Houston, 1990–2020. EPA-COUNCIL-08-001. July. Available at
http://yosemite.epa.gov/sab/sabproduct.nsf/D4D7EC9DAEDA8A548525748600728A83/$File/EPA-COUNCIL-08-001-unsigned.pdf.

U.S. Environmental Protection Agency—Science Advisory Board (U.S. EPA-SAB). 2009b.
Review of EPA's Integrated Science Assessment for Particulate Matter (First External Review Draft, December 2008). EPA-COUNCIL-09-008. May. Available on the Internet at
http://yosemite.epa.gov/sab/SABPRODUCT.NSF/81e39f4c09954fcb85256ead006be86e/73ACCA834AB44A10852575BD0064346B/$File/EPA-CASAC-09-008-unsigned.pdf.

U.S. Environmental Protection Agency—Science Advisory Board (U.S. EPA-SAB). 2009c.
Review of Integrated Science Assessment for Particulate Matter (Second External Review Draft, July 2009). EPA-CASAC-10-001. November. Available on the Internet at
http://yosemite.epa.gov/sab/SABPRODUCT.NSF/81e39f4c09954fcb85256ead006be86e/151B1F83B023145585257678006836B9/$File/EPA-CASAC-10-001-unsigned.pdf.

U.S. Environmental Protection Agency—Science Advisory Board (U.S. EPA-SAB). 2010a.
Review of EPA's DRAFT Health Benefits of the Second Section 812 Prospective Study of the Clean Air Act. EPA-COUNCIL-10-001. June. Available on the Internet at
http://yosemite.epa.gov/sab/sabproduct.nsf/9288428b8eeea4c885257242006935a3/59e06b6c5ca66597852575e7006c5d09!OpenDocument&TableRow=2.3#2.

U.S. Environmental Protection Agency—Science Advisory Board (U.S. EPA-SAB). 2011.
Review of Valuing Mortality Risk Reductions for Environmental Policy: A White Paper (December 10, 2010). EPA-SAB-11-011 July. Available on the Internet at
http://yosemite.epa.gov/sab/sabproduct.nsf/298E1F50F844BC23852578DC0059A616/$File/EPA-SAB-11-011-unsigned.pdf .

SECTION 8
COMPARISON OF MONETIZED BENEFITS AND COSTS

8.1   Summary

Because we are unable to monetize the co-benefits associated with reducing other pollutants such as VOCs and CO, all monetized benefits reflect improvements in ambient $PM_{2.5}$ concentrations. This results in an underestimate of the monetized benefits. Using a 3% discount rate, we estimate the total monetized benefits of this proposed rule to be $1.8 billion to $4.2 billion in the 2014–2022 time frame (Table 8-1). We estimate the impacts for the time frame from 2014 to 2022 in order to provide an average of annualized results for these options from the time of rule promulgation in 2014 to the time of full implementation of both options, which occurs by 2022. The variability of annual impacts for each option provides an appropriate rationale for presenting impacts averaged over this time frame. Using a 7% discount rate, we estimate the total monetized benefits to be $1.7 billion to $3.8 billion in the 2014–2022 time frame. For the Alternative option, using a 3% discount rate, we estimate the total monetized benefits of this proposed rule to be $1.9 billion to $4.2 billion in the 2014–2022 time frame (Table 8-1). Using a 7% discount rate, we estimate the total monetized benefits under the Alternative option to be $1.7 billion to $3.8 billion in the 2014–2022 time frame. The annualized social costs are $15.7 million for the Proposed rule and $28.3 million for the Alternative option in the 2014–2022 time frame (2010 dollars), and are $14.8 million for the proposed rule and $26.9 million for the Alternative option, respectively, in the same time frame using a 3% interest rate. The net benefits (benefits – costs) are therefore $1.8 billion to $4.1 billion at a 3% discount rate for the benefits and $1.7 billion to $3.7 billion at a 7% discount rate for the Proposed option and $1.8 billion to $4.2 Billion at a 3% discount rate and $1.7 billion to $3.8 billion at a 7% discount rate for the Alternative option in the 2014–2022 time frame. The net benefits with annualized social costs at a 3% interest rate are essentially identical to those shown above with costs at a 7% interest rate. Annual benefits were equal through all options thereafter. All estimates are in 2010$. The benefits from reducing other air pollutants have not been monetized in this analysis, including reducing nearly 3,200 tons of VOC, nearly 33,000 tons of CO, black carbon and several HAPs emissions such as formaldehyde and benzene among others each year.

Figure 8-1 shows the full range of net benefits estimates (i.e., annual benefits minus annualized costs) quantified in terms of $PM_{2.5}$ benefits reflecting the average annual impact for the 2014–2022 time frame of the analysis under the Proposed option, and Figure 8-2 shows the full range of net benefits estimates for the Alternative option. The net benefits reflect a 3% discount rate for the benefits.

Table 8-1. Summary of the Monetized Benefits, Social Costs, and Net Benefits for the Proposed Residential Wood Heater NSPS in the 2014–2022 Time Frame ($2010 millions)[a]

Proposed Option	3% Discount Rate			7% Discount Rate		
Total Monetized Benefits[b]	$1,800	to	$4,200	$1,700	to	$3,800
Total Social Costs[c]		$15			$16	
Net Benefits	$1,800	to	$4,100	$1,700	To	$3,700
Nonmonetized Benefits	32,600 tons of CO					
	3,200 tons of VOC					
	Reduced exposure to HAPs, including formaldehyde, benzene, and polycyclic organic matter					
	Reduced Climate effects due to reductions in black carbon emissions					
	Reduced ecosystem effects					
	Reduced visibility impairment					
Alternative Option						
Total Monetized Benefits[b]	$1,900	to	$4,200	$1,700	To	$3,800
Total Social Costs[c]		$27			$28	
Net Benefits	$1,800	to	$4,200	$1,700	to	$3,800
Nonmonetized Benefits	32,900 tons of CO					
	3,200 tons of VOC					
	Reduced exposure to HAPs, including formaldehyde, benzene, and polycyclic organic matter					
	Reduced Climate effects due to reductions in black carbon emissions					
	Reduced ecosystem effects					
	Reduced visibility impairment					

[a] All estimates are for the time frame from 2014 to 2022 inclusive and are rounded to two significant figures. These results include units anticipated to come online and the lowest cost disposal assumption. Total annualized social costs are estimated at a 7% interest rate.

[b] The total monetized benefits reflect the human health benefits associated with reducing exposure to $PM_{2.5}$ through reductions of directly emitted $PM_{2.5}$. It is important to note that the monetized benefits include many but not all health effects associated with $PM_{2.5}$ exposure. Benefits are shown as a range from Krewski et al. (2009) to Lepeule et al. (2012). These models assume that all fine particles, regardless of their chemical composition, are equally potent in causing premature mortality because the scientific evidence is not yet sufficient to allow differentiation of effect estimates by particle type. Because these estimates were generated using benefit-per-ton estimates, we do not break down the total monetized benefits into specific components here. See Figure 7-1 in this RIA for an illustration of the breakdown, or the RIA for the final Cross-States Air Pollution Rule (EPA, 2011) for more information.

[c] The annualized social costs are $14.8 million for the Proposed Option and $26.9 million for the Alternative Option at a 3% interest rate. We assume that annual compliance costs serve as an approximation of the social costs of the proposal.

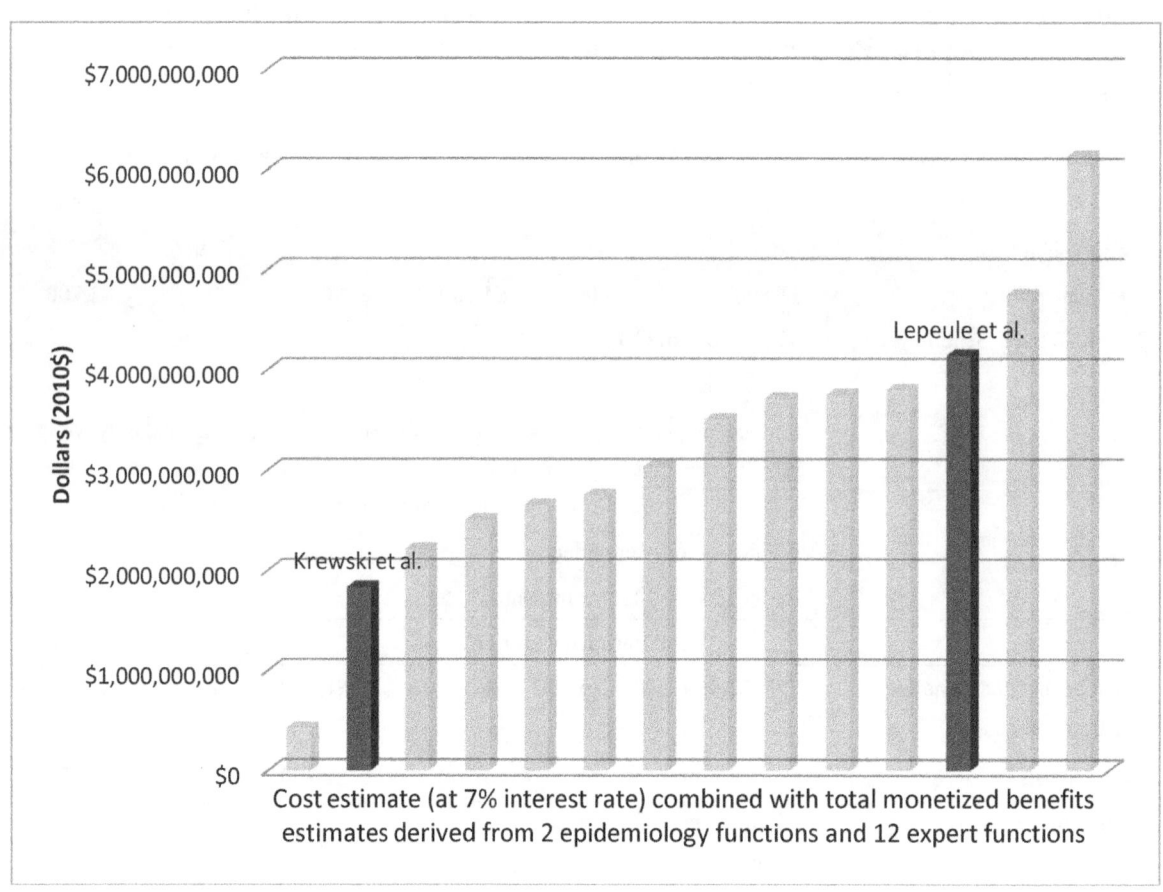

Figure 8-1.   Net Annual Benefits Range in 2014–2022 Time Frame for PM$_{2.5}$ Reductions for the Proposed Option

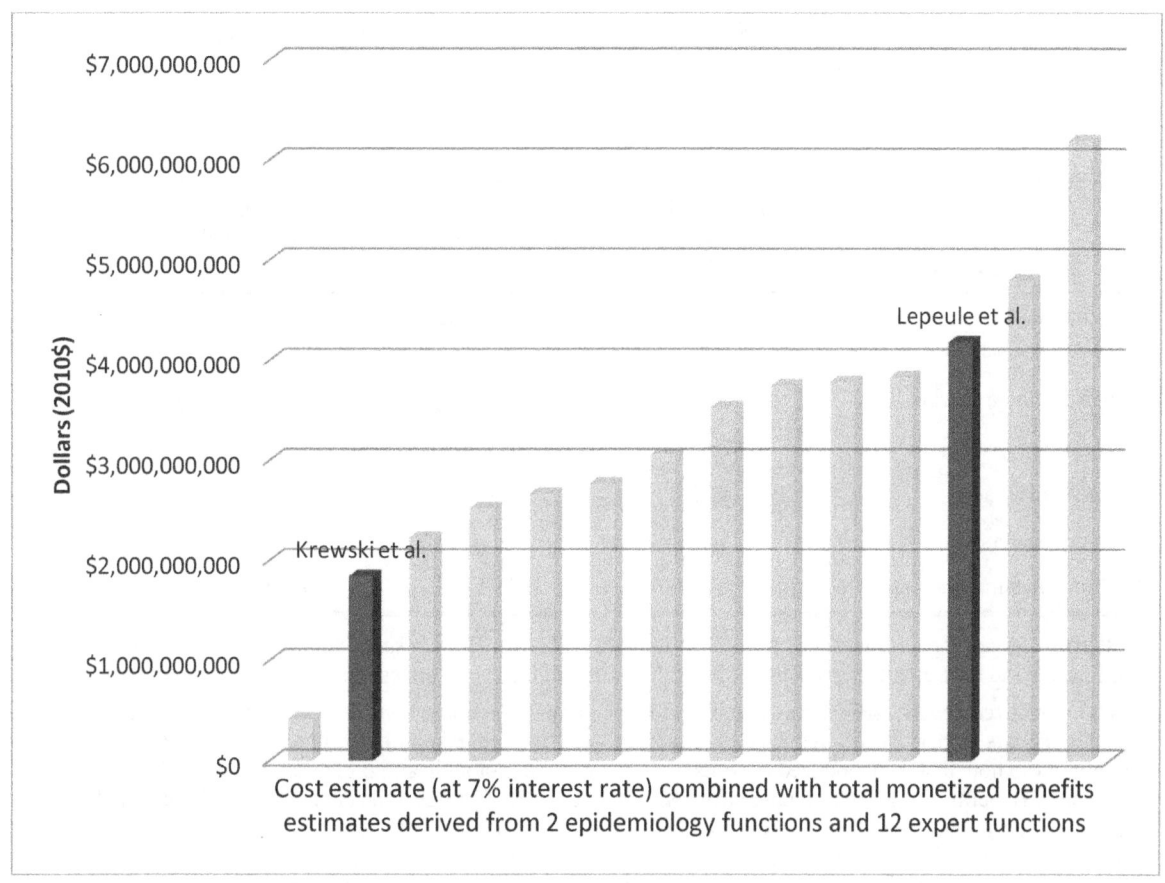

Figure 8-2. Net Annual Benefits Range in 2014–2022 Time Frame for PM$_{2.5}$ Reductions for the Alternative Option

Table 8-2 provides estimates of annualized costs and benefits for each affected source category for the Proposed option, and Table 8-3 provides estimates of annualized costs and benefits for each affected category for the Alternative option.

Table 8-2. Compliance Costs, Monetized Benefits, and Monetized Net Benefits (2010 dollars) by Source Category in the 2014–2022 Time Frame—Proposal Option

Source Category	Total Annualized Costs ($ millions)	Monetized Benefits ($ millions)[a]	Monetized Net Benefits ($ millions)
Wood stoves	$4.2	$62 to $140	$62 to $140
Single burn rate stoves	$0.9	$290 to $650	$290 to $650
Pellet stoves	$3.5	$19 to $43	$19 to $43
Forced-air furnaces	$2.3	$1,000 to $2,200	$1,000 to $2,200
Masonry heaters	$0.3	N/A	N/A
Hydronic heating systems	$4.5	$480 to $1,100	$480 to $1,100

[a] All estimates are for the time frame from 2014 to 2022 inclusive. These results include units anticipated to come online and the lowest cost disposal assumption. Total annualized costs are estimated at a 7% interest rate.

[b] Total monetized benefits are estimated at a 3% discount rate. The total monetized benefits reflect the human health benefits associated with reducing exposure to $PM_{2.5}$ through reductions of directly emitted $PM_{2.5}$. It is important to note that the monetized benefits include many but not all health effects associated with $PM_{2.5}$ exposure. Benefits are shown as a range from Krewski et al. (2009) to Lepeule et al. (2012). These models assume that all fine particles, regardless of their chemical composition, are equally potent in causing premature mortality because the scientific evidence is not yet sufficient to allow differentiation of effect estimates by particle type. Because these estimates were generated using benefit-per-ton estimates, we do not break down the total monetized benefits into specific components here. See Figure7-1 in this RIA for an illustration of the breakdown, or the RIA for the final Cross-States Air Pollution Rule (EPA, 2011) for more information.

Table 8-3. Compliance Costs, Monetized Benefits, and Monetized Net Benefits (2010 dollars) by Source Category in the 2014–2022 Time Frame—Alternative Option

Source Category	Total Annualized Costs ($ millions)[a]	Monetized Benefits ($ millions)[a]	Monetized Net Benefits ($millions)[a]
Wood stoves	$8.1	$52 to $120	$40 to $110
Single burn rate stoves	$1.5	$290 to $650	$290 to $650
Pellet stoves	$6.2	$ $3 to $15	$ $3 to $15
Forced-air furnaces	$3.8	$1,000 to $2,300	$1,000 to $2,300
Masonry heaters	$0.3	N/A	N/A
Hydronic heating systems	$8.3	$470 to $1,100	$470 to $1,100

[a] All estimates are for the time frame from 2014 to 2022 inclusive. These results include units anticipated to come online and the lowest cost disposal assumption. Total annualized costs are estimated at a 7% interest rate.

[b] Total monetized benefits are estimated at a 3% discount rate. The total monetized benefits reflect the human health benefits associated with reducing exposure to $PM_{2.5}$ through reductions of directly emitted $PM_{2.5}$. It is important to note that the monetized benefits include many but not all health effects associated with $PM_{2.5}$ exposure. Benefits are shown as a range from Krewski et al. (2009) to Lepeule et al. (2012). These models assume that all fine particles, regardless of their chemical composition, are equally potent in causing premature mortality because the scientific evidence is not yet sufficient to allow differentiation of effect estimates by particle type. Because these estimates were generated using benefit-per-ton estimates, we do not break down the total monetized benefits into specific components here. See Figure7-1 in this RIA for an illustration of the breakdown, or the RIA for the final Cross-States Air Pollution Rule (EPA, 2011) for more information.

SECTION 9
REFERENCES AND COST APPENDIX

Abt Associates, Inc. 2012. *BenMAP User's Manual Appendices*. Prepared for U.S. Environmental Protection Agency Office of Air Quality Planning and Standards. Research Triangle Park, NC. September. Available on the Internet at http://www.epa.gov/air/benmap/models/BenMAPAppendicesOct2012.pdf.

Agency for Toxic Substances and Disease Registry (ATSDR). 1995. *Toxicological Profile for Polycyclic Aromatic Hydrocarbons (PAHs)*. Atlanta, GA: U.S. Department of Health and Human Services, Public Health Service. Available at: http://www.atsdr.cdc.gov/ToxProfiles/TP.asp?id=122&tid=25

American Enterprise Institute (AEI) and Brookings Institution Joint Center for Regulatory Studies. 1986. *NSPS for Residential Wood Heaters. Regulatory Impact Analysis*. RIN: 2060-AB68. Available at: http://reg-markets.org/admin/authorpdfs/redirect-safely.php?fname=../pdffiles/2060-AB68.RIA.pdf

Bernstein, M.A., and J. Griffin. 2005. *Regional Differences in the Price-Elasticity of Demand for Energy*. The RAND Corporation. Available at: http://www.rand.org/pubs/technical_reports/2005/RAND_TR292.pdf.

Chernov, A. 2008. *Masonry Heaters: Planning Guide for Architects, Home Designers, and builders*. Stovemaster Web site. Available at: http://www.stovemaster.com/files/masonry.pdf.

Dagan, D. January 7, 2005. "Warming Up to Stoves." *Central Penn Business Journal.* Available at: http://www.allbusiness.com/sales/1033217-1.html

Dun & Bradstreet (D&B). 2010. *D&B Marketplace*. A company information database searchable by NAICS code. Accessed on July 15, 2010.

Fann, N., C.M. Fulcher, B.J. Hubbell. 2009. "The influence of location, source, and emission type in estimates of the human health benefits of reducing a ton of air pollution." *Air Qual Atmos Health* 2:169–176.

Fann, N., K.R. Baker, and C.M. Fulcher. 2012. "Characterizing the $PM_{2.5}$-related health benefits of emission reductions for 17 industrial, area and mobile emission sectors across the U.S." *Environment International* 49 41–151.

Gwinn, M.R., J. Craig, D.A. Axelrad, R. Cook, C. Dockins, N. Fann, R. Fegley, D.E. Guinnup, G. Helfand, B. Hubbell, S.L. Mazur, T. Palma, R.L. Smith, J. Vandenberg, and B. Sonawane. 2011. "Meeting report: Estimating the benefits of reducing hazardous air pollutants—summary of 2009 workshop and future considerations." *Environ Health Perspect*. Jan; 119(1): p. 125-30.

Industrial Economics, Inc (IEc). 2009. *Section 812 Prospective Study of the Benefits and Costs of the Clean Air Act: Air Toxics Case Study—Health Benefits of Benzene Reductions in Houston, 1990–2020*. Final Report, July 14, 2009. Available on the Internet at http://www.epa.gov/air/sect812/dec09/812CAAA_Benzene_Houston_Final_Report_July_2009.pdf.

Industrial Economics, Incorporated (IEc). 2006. *Expanded Expert Judgment Assessment of the Concentration-Response Relationship Between $PM_{2.5}$ Exposure and Mortality*. Prepared for: Office of Air Quality Planning and Standards, U.S. Environmental Protection Agency, Research Triangle Park, NC. September. Available on the Internet at http://www.epa.gov/ttn/ecas/regdata/Uncertainty/pm_ee_tsd_expert_interview_summaries.pdf.

Krewski D, Jerrett M, Burnett RT, Ma R, Hughes E, Shi, Y, et al. 2009. *Extended follow-up and spatial analysis of the American Cancer Society study linking particulate air pollution and mortality.* HEI Research Report, 140, Health Effects Institute, Boston, MA.

Lepeule J, Laden F, Dockery D, Schwartz J 2012. "Chronic Exposure to Fine Particles and Mortality: An Extended Follow-Up of the Harvard Six Cities Study from 1974 to 2009." *Environ Health Perspect.* Jul;120(7):965-70.

National Research Council (NRC). 2002. *Estimating the Public Health Benefits of Proposed Air Pollution Regulations.* Washington, DC: The National Academies Press.

Office of Management and Budget (OMB). 2003. *Circular A-4: Regulatory Analysis.* Washington, DC. Available on the Internet at http://www.whitehouse.gov/omb/circulars/a004/a-4.html.

Roman, Henry A., Katherine D. Walker, Tyra L. Walsh, Lisa Conner, Harvey M. Richmond, Bryan J. Hubbell, and Patrick L. Kinney. 2008. "Expert Judgment Assessment of the Mortality Impact of Changes in Ambient Fine Particulate Matter in the U.S." *Environ. Sci. Technol.*, 42(7):2268-2274.

Sisler, J.F. 1996. *Spatial and seasonal patterns and long-term variability of the composition of the haze in the United States: an analysis of data from the IMPROVE network.* CIRA Report, ISSN 0737-5352-32, Colorado State University.

U.S. Census Bureau. 2008a. Firm Size Data from the Statistics of U.S. Businesses: U.S. Detail Employment Sizes: 2002. <http://www2.census.gov/csd/susb/2002/ 02us_detailed%20sizes_6digitnaics.txt >

U.S. Environmental Protection Agency—Science Advisory Board (U.S. EPA-SAB). 2002. *Workshop on the Benefits of Reductions in Exposure to Hazardous Air Pollutants: Developing Best Estimates of Dose-Response Functions An SAB Workshop Report of an EPA/SAB Workshop (Final Report).* EPA-SAB-EC-WKSHP-02-001. January. Available on the Internet at http://yosemite.epa.gov/sab%5CSABPRODUCT.NSF/34355712EC011A358525719A005BF6F6/$File/ecwkshp02001%2Bappa-g.pdf.

U.S. Environmental Protection Agency (U.S. EPA). 1995. *Regulatory Impact Analysis for the Petroleum Refinery NESHAP. Revised Draft for Promulgation.* Office of Air Quality Planning and Standards, Research Triangle Park, N.C. Available on the Internet at http://www.reg-markets.org/admin/authorpdfs/page.php?id=705.

U.S. Environmental Protection Agency (U.S. EPA). 2006. *Air Quality Criteria for Ozone and Related Photochemical Oxidants (Final).* EPA/600/R-05/004aF-cF. Washington, DC: U.S. EPA. February. Available on the Internet at http://cfpub.epa.gov/ncea/CFM/recordisplay.cfm?deid=149923.

U.S. Environmental Protection Agency (U.S. EPA). 2008a. *Integrated Science Assessment for Sulfur Oxides—Health Criteria (Final Report).* National Center for Environmental Assessment, Research Triangle Park, NC. September. Available on the Internet at http://cfpub.epa.gov/ncea/cfm/recordisplay.cfm?deid=198843.

U.S. Environmental Protection Agency (U.S. EPA). 2008c. *Integrated Science Assessment for Oxides of Nitrogen—Health Criteria (Final Report).* National Center for Environmental Assessment, Research Triangle Park, NC. July. Available at at http://cfpub.epa.gov/ncea/cfm/recordisplay.cfm?deid=194645.

U.S. Environmental Protection Agency (U.S. EPA). 2008d. *Integrated Science Assessment for Oxides of Nitrogen and Sulfur–Ecological Criteria National (Final Report)*. National Center for Environmental Assessment, Research Triangle Park, NC. EPA/600/R-08/139. December. Available at http://cfpub.epa.gov/ncea/cfm/recordisplay.cfm?deid=201485.

U.S. Environmental Protection Agency (U.S. EPA). 2009. *Integrated Science Assessment for Particulate Matter (Final Report)*. EPA-600-R-08-139F. National Center for Environmental Assessment—RTP Division. December. Available on the Internet at http://cfpub.epa.gov/ncea/cfm/recordisplay.cfm?deid=216546.

U.S. Environmental Protection Agency (U.S. EPA). 2010a. *Integrated Science Assessment for Carbon Monoxide*. National Center for Environmental Assessment, Research Triangle Park, NC. EPA/600/R-09/019F. January. Available at http://cfpub.epa.gov/ncea/cfm/recordisplay.cfm?deid=218686.

U.S. Environmental Protection Agency (U.S. EPA). 2010b. *Technical Support Document: Summary of Expert Opinions on the Existence of a Threshold in the Concentration-Response Function for $PM_{2.5}$-related Mortality*. Research Triangle Park, NC. June. Available on the Internet at www.epa.gov/ttn/ecas/regdata/Benefits/thresholdstsd.pdf.

U.S. Environmental Protection Agency (U.S. EPA). 2010c. *Valuing Mortality Risk Reductions for Environmental Policy: A White Paper: SAB Review Draft*. National Center for Environmental Economics December. Available on the Internet at http://yosemite.epa.gov/ee/epa/eerm.nsf/vwAN/EE-0563-1.pdf/$file/EE-0563-1.pdf.

U.S. Environmental Protection Agency (U.S. EPA). 2010e. *Guidelines for Preparing Economic Analyses*. EPA 240-R-10-001. National Center for Environmental Economics, Office of Policy Economics and Innovation. Washington, DC. December. Available on the Internet at http://yosemite.epa.gov/ee/epa/eerm.nsf/vwAN/EE-0568-50.pdf/$file/EE-0568- 50.pdf.

U.S. Environmental Protection Agency (U.S. EPA). 2011a. *The Benefits and Costs of the Clean Air Act from 1990 to 2020*. Office of Air and Radiation, Washington, DC. March. http://www.epa.gov/air/sect812/feb11/fullreport.pdf. Accessed March 30, 2011.

U.S. Environmental Protection Agency (U.S. EPA). 2011b. *Regulatory Impact Analysis for the Final Mercury and Air Toxics Standards*. EPA-452/R-11-011. December. Available on the Internet at http://www.epa.gov/ttn/ecas/regdata/RIAs/matsriafinal.pdf.

U.S. Environmental Protection Agency (U.S. EPA). 2011c. *2005 National-Scale Air Toxics Assessment.* Office of Air Quality Planning and Standards, Research Triangle Park, NC. March. Available on the Internet at http://www.epa.gov/ttn/atw/nata2005/.

U.S. Environmental Protection Agency (U.S. EPA). 2011d. *Regulatory Impact Analysis: National Emission Standards for Hazardous Air Pollutants for Industrial, Commercial, and Institutional Boilers and Process Heaters*. February. Available on the Internet at http://www.epa.gov/ttnecas1/regdata/RIAs/boilersriafinal110221_psg.pdf.

U.S. Environmental Protection Agency (U.S. EPA). 2012a. *Regulatory Impact Analysis for the Final Revisions to the National Ambient Air Quality Standards for Particulate Matter*. EPA-452/R-12-003. Office of Air Quality Planning and Standards, Health and Environmental Impacts Division. December. Available at http://www.epa.gov/pm/2012/finalria.pdf.

U.S. Environmental Protection Agency (U.S. EPA). 2012b. *Regulatory Impact Analysis: Petroleum Refineries New Source Performance Standards Ja.* Office of Air Quality Planning and Standards, Health and Environmental Impacts Division. June. Available at http://www.epa.gov/ttnecas1/regdata/RIAs/refineries_nsps_ja_final_ria.pdf.

U.S. Environmental Protection Agency (U.S. EPA). 2013. *Technical Support Document: Estimating the Benefit per ton of Reducing PM$_{2.5}$ Precursors from 17 sectors.* Office of Air Quality Planning and Standards, Research Triangle Park, NC. February.

U.S. Environmental Protection Agency—Science Advisory Board (U.S. EPA-SAB). 2008. *Characterizing Uncertainty in Particulate Matter Benefits Using Expert Elicitation.* EPA-COUNCIL-08-002. July. Available on the Internet at http://yosemite.epa.gov/sab/sabproduct.nsf/0/43B91173651AED9E85257487004EA6CB/$File/EPA-COUNCIL-08-002-unsigned.pdf.

U.S. Environmental Protection Agency—Science Advisory Board (U.S. EPA-SAB). 2000. *An SAB Report on EPA's White Paper Valuing the Benefits of Fatal Cancer Risk Reduction.* EPA-SAB-EEAC-00-013. July. Available on the Internet at http://yosemite.epa.gov/sab%5CSABPRODUCT.NSF/41334524148BCCD6852571A700516498/$File/eeacf013.pdf.

U.S. Environmental Protection Agency—Science Advisory Board (U.S. EPA-SAB). 2004c. *Advisory Council on Clean Air Compliance Analysis Response to Agency Request on Cessation Lag.* EPA-COUNCIL-LTR-05-001. December. Available on the Internet at <. http://yosemite.epa.gov/sab/sabproduct.nsf/0/39F44B098DB49F3C85257170005293E0/$File/council_ltr_05_001.pdf

U.S. Environmental Protection Agency—Science Advisory Board (U.S. EPA-SAB). 2008. *Benefits of Reducing Benzene Emissions in Houston, 1990–2020.* EPA-COUNCIL-08-001. July. Available at http://yosemite.epa.gov/sab/sabproduct.nsf/D4D7EC9DAEDA8A548525748600728A83/$File/EPA-COUNCIL-08-001-unsigned.pdf.

U.S. Environmental Protection Agency—Science Advisory Board (U.S. EPA-SAB). 2009b. *Review of EPA's Integrated Science Assessment for Particulate Matter (First External Review Draft, December 2008).* EPA-COUNCIL-09-008. May. Available on the Internet at http://yosemite.epa.gov/sab/SABPRODUCT.NSF/81e39f4c09954fcb85256ead006be86e/73ACCA834AB44A10852575BD0064346B/$File/EPA-CASAC-09-008-unsigned.pdf.

U.S. Environmental Protection Agency—Science Advisory Board (U.S. EPA-SAB). 2009c. *Review of Integrated Science Assessment for Particulate Matter (Second External Review Draft, July 2009).* EPA-CASAC-10-001. November. Available on the Internet at http://yosemite.epa.gov/sab/SABPRODUCT.NSF/81e39f4c09954fcb85256ead006be86e/151B1F83B023145585257678006836B9/$File/EPA-CASAC-10-001-unsigned.pdf.

U.S. Environmental Protection Agency—Science Advisory Board (U.S. EPA-SAB). 2010a. *Review of EPA's DRAFT Health Benefits of the Second Section 812 Prospective Study of the Clean Air Act.* EPA-COUNCIL-10-001. June. Available on the Internet at http://yosemite.epa.gov/sab/sabproduct.nsf/9288428b8eeea4c885257242006935a3/59e06b6c5ca66597852575e7006c5d09!OpenDocument&TableRow=2.3#2.

U.S. Environmental Protection Agency—Science Advisory Board (U.S. EPA-SAB). 2011. *Review of Valuing Mortality Risk Reductions for Environmental Policy: A White Paper (December 10, 2010).* EPA-SAB-11-011 July. Available on the Internet at http://yosemite.epa.gov/sab/sabproduct.nsf/298E1F50F844BC23852578DC0059A616/$File/EPA-SAB-11-011-unsigned.pdf.

Fann, N., C. M. Fulcher, and B. J. Hubbell. 2009. "The Influence of Location, Source, and Emission Type in Estimates of the Human Health Benefits of Reducing a Ton of Air Pollution." *Air Quality, Atmosphere, and Health* 2:169-176.

Fireplaces & Woodstoves. 2010. "Masonry Heaters." http://www.fireplacesandwoodstoves.com/indoor-fireplaces/masonry-fireplaces.aspx.

Frost & Sullivan. 2010. *Project: Market Research and Report on North American Residential Wood Heaters, Fireplaces, and Hearth Heating Products Market* (P.O. # PO1-IMP402-F&S). Prepared for EC/R.

Fullerton, D., and G. Metcalf. 2002. "Tax Incidence." In A. Auerbach and M. Feldstein, eds., Handbook of Public Economics, Vol. 4, Amsterdam: Elsevier.Hearth, Patio, and Barbeque Association (HPBA). 2010a. "Comments for the Small Business Advocacy Review (SBAR) Panel, Regarding the Revision of New Source Performance Standards for New Residential Wood Heaters." Comments submitted to EPA on September 12, 2010.

Gwinn, M.R., J. Craig, D.A. Axelrad, R. Cook, C. Dockins, N. Fann, R. Fegley, D.E. Guinnup, G. Helfand, B. Hubbell, S.L. Mazur, T. Palma, R.L. Smith, J. Vandenberg, and B. Sonawane. 2011. "Meeting report: Estimating the benefits of reducing hazardous air pollutants—summary of 2009 workshop and future considerations." *Environ Health Perspect.* 119(1): p. 125-30.

Hearth, Patio, and Barbeque Association (HPBA). 2010b. "Fireplace Insert Fact Sheet." Available at: http://static.hpba.org/fileadmin/factsheets/product/FS_FireplaceInsert.pdf.

Hearth, Patio, and Barbecue Association (HPBA). 2010c. "Outdoor Heating Options." Available at: http://www.hpba.org/consumers/outdoor-living/outdoor-heating-options.

Houck, J. 2009. "Let's Not Forget about Coal." *Hearth & Home.* December. Available at: http://www.hearthandhome.com/articles.html.

Houck, J., and P. Tiegs. 2009. "There's a Freight Train Comin'." *Hearth & Home.* December. Available at: http://www.hearthandhome.com/articles.html.

Industrial Economics, Inc (IEc). 2006. *Expanded Expert Judgment Assessment of the Concentration-Response Relationship Between PM$_{2.5}$ Exposure and Mortality.* Prepared for the U.S. EPA, Office of Air Quality Planning and Standards, September. Available at: http://www.epa.gov/ttn/ecas/regdata/Uncertainty/pm_ee_report.pdf.

Industrial Economics, Inc (IEc). 2009. *Section 812 Prospective Study of the Benefits and Costs of the Clean Air Act: Air Toxics Case Study—Health Benefits of Benzene Reductions in Houston, 1990–2020.* Final Report, July 14, 2009. http://www.epa.gov/air/sect812/dec09/812CAAA+Benzene_Houston_Final_Report_July_2009.pdf. Accessed March 30, 2011.

Kochi, I., B. Hubbell, and R. Kramer. 2006. "An Empirical Bayes Approach to Combining Estimates of the Value of Statistical Life for Environmental Policy Analysis." *Environmental and Resource Economics* 34:385-406.

Krewski D, Jerrett M, Burnett RT, Ma R, Hughes E, Shi, Y, et al. 2009. *Extended follow-up and spatial analysis of the American Cancer Society study linking particulate air pollution and mortality.* HEI Research Report, 140, Health Effects Institute, Boston, MA.

Laden, F., J. Schwartz, F.E. Speizer, and D.W. Dockery. 2006. Reduction in Fine Particulate Air Pollution and Mortality. *American Journal of Respiratory and Critical Care Medicine.* 173: 667-672.

Landsburg, S.H. 2005. *Price Theory and Applications. 6th Ed.* Thomson South Western.Mason, OH.

Mankiw, N.G. 1998. *Principles of Economics.* Orlando, Fl: Dryden Press.

Masonry Heater Association of North America (MHA). 1998. MHA Masonry Heater Definition. Available at: http://mha-net.org/docs/def-mha.htm.

Morgenstern, R. D., W. A. Pizer, and J. S. Shih. 2002. "Jobs versus the Environment: An Industry-Level Perspective." *Journal of Environmental Economics and Management* 43(3):412-436. Available at http://ac.els-cdn.com/S009506960191191X/1-s2.0-S009506960191191X-main.pdf?_tid=6bc8845e-7c56-11e2-84a7-00000aab0f26&acdnat=1361472319_4fe7ef315e9b2f8fc064c0a767895205 (subscription required).

Mrozek, J.R., and L.O. Taylor. 2002. "What Determines the Value of Life? A Meta-Analysis." Journal of Policy Analysis and Management 21(2):253-270.

National Research Council (NRC). 2002. *Estimating the Public Health Benefits of Proposed Air Pollution Regulations.* Washington, DC: The National Academies Press.

Nicholson, Walter. 1998. *Microeconomic Theory*. Orlando: The Dryden Press.

Northeast States for Coordinated Air Use Management (NESCAUM). 2006. *Assessment of Outdoor Wood-Fired Boilers.* March, 2006 (Revised June, 2006). Available at: http://www.nescaum.org/documents/assessment-of-outdoor-wood-fired-boilers.

Office of Management and Budget (OMB). 2003. *Circular A-4: Regulatory Analysis.* Washington, DC. Available on the Internet at http://www.whitehouse.gov/omb/circulars/a004/a-4.html.

Pope, C.A., III, R.T. Burnett, M.J. Thun, E.E. Calle, D. Krewski, K. Ito, and G.D. Thurston. 2002. "Lung Cancer, Cardiopulmonary Mortality, and Long-term Exposure to Fine Particulate Air Pollution." *Journal of the American Medical Association* 287:1132-1141.

Roman, Henry A., Katherine D. Walker, Tyra L. Walsh, Lisa Conner, Harvey M. Richmond, Bryan.J. Hubbell, and Patrick L. Kinney. 2008. "Expert Judgment Assessment of the Mortality Impact of Changes in Ambient Fine Particulate Matter in the U.S. Environmental Science & Technology 42(7):2268-2274.

Seaton, T. 2010. "Masonry Heater Industry Analysis: Residential Sold Fuel Burning Appliance SBREFA Process." Industry comments submitted to EPA on September 12, 2010.

Sisler, J.F. 1996. *Spatial and seasonal patterns and long-term variability of the composition of the haze in the United States: an analysis of data from the IMPROVE network*. CIRA Report, ISSN 0737-5352-32, Colorado State University.

The Risk Management Association. 2008. *Annual Statement Studies, Financial Ratio Benchmarks 2008–2009*. Risk Management Association, Philadelphia: 2008.

U.S. Census Bureau. 2007. *Survey of Plant Capacity: 2006*. "Table 1a. Full Capacity Utilization Rates by Industry Fourth Quarter 2002–2006." U.S. Census Bureau: Washington DC. Report No. MQ-C1(06). Available at: http://www.census.gov/manufacturing/capacity/historical_data/index.html.

U.S. Census Bureau. 2009. *American Community Survey: 2006–2008*. Available at: http://factfinder.census.gov/servlet/DatasetMainPageServlet?_program=ACS&submenId=&_lang+en&_ts=. U.S. Census Bureau. 2010a. American Fact Finder. Sector 31: Annual Survey of Manufactures: General Statistics: Statistics for Industry Groups and Industries: 2008 and 2007. Available at: http://factfinder.census.gov. Accessed July 20, 2010.

U.S. Census Bureau. 2010b. American Fact Finder. Sector 23: EC0723SG01: Construction: Summary Series: General Summary: Detailed Statistics for Establishments: 2007. Available at: http://factfinder.census.gov. Accessed July 20, 2010.

U.S. Census Bureau. 2010c. American Fact Finder. Sector 42: EC0742A1: Wholesale Trade: Geographic Area Series: Summary Statistics for the United States, States, Metro Areas, Counties, and Places: 2007. Available at: http://factfinder.census.gov. Accessed July 20, 2010.

U.S. Census Bureau. 2010d. American Fact Finder. Sector 44: EC0744A1: Retail Trade: Geographic Area Series: Summary Statistics for the United States, States, Metro Areas, Counties, and Places: 2007. Available at: http://factfinder.census.gov. Accessed July 20, 2010.

U.S. Census Bureau. 2010e. Census Regions and Divisions of the United States. Available at: http://www.census.gov/geo/www/us_regdiv.pdf. Accessed September 12, 2010.

U.S. Census Bureau. 2010f. North American Industrial Classification System [NAICS] Code Definitions Web site. Available at: http://www.census.gov/eos/www/naics/.

U.S. Department of Energy (DOE). 2009. "Your Home: Selecting Heating Fuel and System Types." ENERGYSTAR Web site. U.S. Department of Energy. Available at: http://www.energysavers.gov/your_home/space_heating_cooling/index.cfm/mytopic=12330..

U.S. Department of Energy (DOE). 2010. "Your Home: Masonry Heaters." ENERGYSTAR Web site. U.S. Department of Energy. Available at: http://www.energysavers.gov/your_home/space_heating_cooling/index.cfm/mytopic=12570.

U.S. Energy Information Administration (EIA). 2011. *Residential Energy Consumption Survey: 2009*. Available at: http://www.eia.doe.gov/emeu/recs/recspubuse05/pubuse05.html.

U.S. Environmental Protection Agency—Science Advisory Board (U.S. EPA-SAB). 2007. *SAB Advisory on EPA's Issues in Valuing Mortality Risk Reduction*. EPA-SAB-08-001. October. Available on the Internet at <http://yosemite.epa.gov/sab/sabproduct.nsf/4128007E7876B8F0852573760058A978/$File/sab-08-001.pdf>.

U.S. Environmental Protection Agency—Science Advisory Board (U.S. EPA-SAB). 2009a. *Review of EPA's Integrated Science Assessment for Particulate Matter (First External Review Draft, December 2008)*. EPA-COUNCIL-09-008. May. Available on the Internet at <http://yosemite.epa.gov/sab/SABPRODUCT.NSF/81e39f4c09954fcb85256ead006be86e/73ACCA834AB44A10852575BD0064346B/$File/EPA-CASAC-09-008-unsigned.pdf>.

U.S. Environmental Protection Agency—Science Advisory Board (U.S. EPA-SAB). 2009b. Consultation on EPA's Particulate Matter National Ambient Air Quality Standards: Scope and Methods Plan for Health Risk and Exposure Assessment. EPA-COUNCIL-09-009. May. Available on the Internet at <http://yosemite.epa.gov/sab/SABPRODUCT.NSF/81e39f4c09954fcb85256ead006be86e/723FE644C5D758DF852575BD00763A32/$File/EPA-CASAC-09-009-unsigned.pdf>.

U.S. Environmental Protection Agency—Science Advisory Board (U.S. EPA-SAB). 2010. *Review of EPA's DRAFT Health Benefits of the Second Section 812 Prospective Study of the Clean Air Act. EPA-COUNCIL-10-001*. June. Available on the Internet at http://yosemite.epa.gov/sab/sabproduct.nsf/0/72D4EFA39E48CDB28525774500738776/$File/EPA-COUNCIL-10-001-unsigned.pdf.

U.S. Environmental Protection Agency (EPA). 2002. "Profile of the Pulp and Paper Industry." *Sector Notebook Project*. Available at: http://www.epa.gov/Compliance/resources/publications/assistance/sectors/notebooks/index.html.

U.S. Environmental Protection Agency (EPA). 2006a. *Final Guidance EPA Rulewriters: Regulatory Flexibility Act as Amended by the Small Business and Regulatory Enforcement Fairness Act*. http://www.epa.gov/sbrefa/documents/rfaguidance11-00-06.pdf.

U.S. Environmental Protection Agency (EPA). 2006b. *Regulatory Impact Analysis. 2006 National Ambient Air Quality Standards for Particulate Matter*. Chapter 5. Available at: http://www.epa.gov/ttn/ecas/regdata/RIAs/Chapter%205--Benefits.pdf.

U.S. Environmental Protection Agency (EPA). 2008a. *Integrated Science Assessment for Sulfur Oxides—Health Criteria* (Final Report). National Center for Environmental Assessment, Research Triangle Park, NC. Available at: http://cfpub.epa.gov/ncea/cfm/recordisplay.cfm?deid=198843.

U.S. Environmental Protection Agency (U.S. EPA). 2008b. *Integrated Science Assessment for Sulfur Oxides—Health Criteria* (Final Report). National Center for Environmental Assessment, Research Triangle Park, NC. Available at http://cfpub.epa.gov/ncea/cfm/recordisplay.cfm?deid=198843.

U.S. Environmental Protection Agency (U.S. EPA). 2008c. *Integrated Science Assessment for Oxides of Nitrogen—Health Criteria (Final Report)*. National Center for Environmental Assessment, Research Triangle Park, NC. July. Available at at http://cfpub.epa.gov/ncea/cfm/recordisplay.cfm?deid=194645.

U.S. Environmental Protection Agency (U.S. EPA). 2008d. *Integrated Science Assessment for Oxides of Nitrogen and Sulfur–Ecological Criteria National* (Final Report). National Center for Environmental Assessment, Research Triangle Park, NC. EPA/600/R-08/139. December. Available at http://cfpub.epa.gov/ncea/cfm/recordisplay.cfm?deid=201485.

U.S. Environmental Protection Agency (U.S. EPA). 2009a. *Regulatory Impact Analysis: National Emission Standards for Hazardous Air Pollutants from the Portland Cement Manufacturing Industry*. Office of Air Quality Planning and Standards, Research Triangle Park, NC. April. Available on the Internet at <http://www.epa.gov/ttn/ecas/regdata/RIAs/portlandcementria_4-20-09.pdf >.

U.S. Environmental Protection Agency (U.S. EPA). 2009b. *Integrated Science Assessment for Particulate Matter (Final Report)*. EPA-600-R-08-139F. National Center for Environmental Assessment—RTP Division. December. Available on the Internet at <http://cfpub.epa.gov/ncea/cfm/recordisplay.cfm?deid=216546>.

U.S. Environmental Protection Agency (EPA). 2009c. *Subpart AAA-Standards of Performance for New Residential Wood Heaters*. Discussion Draft. Available at: http://www.hpba.org/admin/NSPS-Review-Document.pdf.

U.S. Environmental Protection Agency (EPA). 2010a. *Integrated Science Assessment for Carbon Monoxide*. National Center for Environmental Assessment, Research Triangle Park, NC. EPA/600/R-09/019F. January. Available at: http://cfpub.epa.gov/ncea/cfm/recordisplay.cfm?deid=218686.

U.S. Environmental Protection Agency (EPA). 2010b. *Final Regulatory Impact Analysis (RIA) for the SO$_2$ National Ambient Air Quality Standards (NAAQS)*. Office of Air Quality Planning and Standards, Research Triangle Park, NC. June. Available at: http://www.epa.gov/ttn/ecas/regdata/RIAs/fso2ria100602full.pdf.

U.S. Environmental Protection Agency (EPA). 2010c. *Guidelines for Preparing Economic Analyses*. EPA 240-R-10-001. Washington, DC: National Center for Environmental Economics, Office of Policy Economics and Innovation. Available at: http://yosemite.epa.gov/ee/epa/eed.nsf/pages/guidelines.html.

U.S. Environmental Protection Agency (EPA). 2010d. *Lowest Measured Level (LML) Assessment for Rules without Policy-Specific Air Quality Data Available: Technical Support Document*. Office of Air Quality Planning and Standards, Research Triangle Park, NC. July. Available at: http://www.epa.gov/ttn/ecas/regdata/Benefits/thresholdstsd.pdf.

U.S. Environmental Protection Agency (EPA). 2010e. *Regulatory Impact Analysis for the Proposed Federal Transport Rule*. Office of Air Quality Planning and Standards, Research Triangle Park, NC. July. Available at: http://www.epa.gov/ttn/ecas/regdata/RIAs/proposaltrria_final.pdf.

U.S. Environmental Protection Agency (EPA). 2010f. *Regulatory Impact Analysis for the SO$_2$ NAAQS*. Office of Air Quality Planning and Standards, Research Triangle Park, NC. June. Available at: http://www.epa.gov/ttn/ecas/regdata/RIAs/fso2ria100602full.pdf.

U.S. Environmental Protection Agency (EPA). May 2004. Final Regulatory Analysis: Control of Emissions from Nonroad Diesel Engines. EPA420-R-04-007. Washington, DC: EPA. http://www.epa.gov/nonroad-diesel/2004fr/420r04007.pdf.

U.S. Environmental Protection Agency (U.S. EPA). 2010b. Technical Support Document: Summary of Expert Opinions on the Existence of a Threshold in the Concentration-Response Function for PM$_{2.5}$-related Mortality. Research Triangle Park, NC. June. Available on the Internet at www.epa.gov/ttn/ecas/regdata/Benefits/thresholdstsd.pdf.

U.S. Environmental Protection Agency (U.S. EPA). 2010c. *Guidelines for Preparing Economic Analyses*. EPA 240-R-10-001. National Center for Environmental Economics, Office of Policy Economics and Innovation. Washington, DC. December. Available on the Internet at <http://yosemite.epa.gov/ee/epa/eerm.nsf/vwAN/EE-0568-50.pdf/$file/EE-0568-50.pdf>.

U.S. Environmental Protection Agency (U.S. EPA). 2011a. *The Benefits and Costs of the Clean Air Act from 1990 to 2020*. Office of Air and Radiation, Washington, DC. March. http://www.epa.gov/air/sect812/feb11/fullreport.pdf. Accessed March 30, 2011.

U.S. Environmental Protection Agency (U.S. EPA). 2011b. *Regulatory Impact Analysis for the Final Transport Rule*. Office of Air Quality Planning and Standards, Research Triangle Park, NC. June. Available at http://www.epa.gov/airtransport/pdfs/FinalRIA.pdf.

U.S. Environmental Protection Agency. 2006. *Regulatory Impact Analysis, 2006 National Ambient Air Quality Standards for Particulate Matter*, Chapter 5. Available at <http://www.epa.gov/ttn/ecas/regdata/RIAs/Chapter%205--Benefits.pdf>.

U.S. Environmental Protection Agency. 2008a. *Regulatory Impact Analysis, 2008 National Ambient Air Quality Standards for Ground-level Ozone*, Chapter 6. Available at <http://www.epa.gov/ttn/ecas/regdata/RIAs/6-ozoneriachapter6.pdf>.

U.S. Housing and Urban Development (HUD). 2008. *American Housing Survey for the United States.* Multiple Years. Table 1A-5. Available at: http://www.census.gov/hhes/www/housing/ahs/nationaldata.html.

U.S. Small Business Administration (SBA), Office of Advocacy. May 2012. A Guide for Government Agencies, How to Comply with the Regulatory Flexibility Act, Implementing the President's Small Business Agenda and Executive Order 13272. Available at http://www.sba.gov/sites/default/files/rfaguide_0512_0.pdf.

U.S. Small Business Administration (SBA). 2013. Table of Small Business Size Standards Matched to North American Industry Classification System Codes. Effective July 22. 2013. http://www.sba.gov/sites/default/files/files/Size_Standards_Table(1).pdf.

Viscusi, V.K., and J.E. Aldy. 2003. "The Value of a Statistical Life: A Critical Review of Market Estimates throughout the World." *Journal of Risk and Uncertainty* 27(1):5-76.

Wade, S.H. 2003. "Price Responsiveness in the AEO2003 NEMS Residential and Commercial Buildings Sector Models." http://www.eia.doe.gov/oiaf/analysispaper/elasticity /pdf/buildings.pdf

Wood Heat Organization. 2010. "Fireplace Inserts: The Cure for Cold Fireplaces." Available at: http://www.woodheat.org/technology/inserts.htm.

APPENDIX

Documentation of Costs for Residential Wood Heaters NSPS Proposal

This appendix of the RIA documents the estimated nationwide cost impacts on manufacturers of emission reduction options being considered for residential wood heaters as part of the New Source Performance Standards (NSPS) review of residential wood heaters. The underlying cost assumptions for two options are summarized herein ▯▯▯▯▯▯▯▯▯▯ ▯▯▯"▯▯▯▯▯"▯▯▯ ▯▯▯ ▯▯▯▯▯▯▯▯▯▯▯▯▯▯"▯▯▯▯▯▯▯▯" ▯▯ which differ in the number of stepped emission limits and in the phased-in compliance dates.

I. Estimated Research and Development (R&D) Costs

A. Residential Wood Heaters – Room Heaters & Central Heaters

We have heard various estimates of the costs to bring a wood heater from conception to completion, from $300,000 for a single model to $1,360,000 for a 4-firebox model line. A recent Hearth and Home article estimated the total cost to bring a model from conception to market as $645,000 to $750,000 for steel stoves and over $1 million for cast-iron, enameled wood stoves. The authors indicated that costs would decrease for separate models in the same line by up to 25%. Based on this information, we estimate that a 4-model steel line would cost up to $328,125 per model to develop. These costs include marketing, design, developing first generation, second generation and prototype units; NSPS and safety testing, equipment tooling, etc. The manufacturer supplying these figures for the article estimated that the NSPS and safety testing component of these costs would constitute $40,000 per model. This manufacturer said that development time is 12 to 14 months for non-catalytic heaters and 10 to 12 months for catalytic heaters.[104]

Another manufacturer estimated costs of new product development, including design, prototype development, testing, tooling equipment and other manufacturing changes, marketing support, materials, training, and education to be in excess of $300,000 over an 8- to 12-month schedule for a relatively uncomplicated product. Costs will increase for products that have more sophisticated controls.[105] One other manufacturer estimated that their typical model development costs are around $200,000/model.[106]

Two manufacturers suggested a 14- to 18-month timeframe is required to develop a new firebox, but added that it will take from 5 to 6 years of intensive engineering and R&D efforts to have a model line consisting of 4 boxes ready-for-manufacture. They agreed that knowledge of the process obtained during each

---

[104] James E. Houck and Paul Tiegs. *There's a Freight Train Comin'*. Hearth and Home. December 2009.
[105] Comments from United States Stove Company, Small Entity Representative. July 13, 2010.
[106] Confidential Business Information.

firebox development will shorten (somewhat) the time necessary, but not enough to consider within a guiding framework. These manufacturers also provided estimated development costs for a 4-box model line, presented in Table A-1.[107]

Table A-1. Example of Manufacturers' Estimates of Costs to Develop Model Line (4 Fireboxes)

Cost Component	Estimated Costs	Notes
Salaries	$850,000	Using 2-full time experienced employees to bring the products to market, salaries and benefits are estimated at $160,000 per year for at least 5.3 years. Tasks include design, prototyping, testing, production-line integration, and marketing.
Laboratory Equipment	$50,000	In order to accelerate R&D and avoid validating each result with independent testing labs (too costly for most manufacturers), new testing equipment will need to be purchased in order to sample flue gases, measure test load weight loss, record data automatically, and analyze flue gas composition.
Prototypes	$25,000	Numerous prototypes will be needed until the final product can be approved. For each firebox, an estimated 8 prototypes will be needed, at a cost of $700 each. Numerous samples of various components will also have to be purchased from vendors.
Test Fuel*	$45,000*	Each test costs at least $50 in fuel (assuming cribs are used), including waste. And estimated 150 tests will have to be conducted for each firebox for a total of $7,500, or $30,000 for a 4-firebox model line based on crib testing.*
Testing Services*	$150,000*	Testing services for emissions, efficiency, and safety are estimated to last approximately 3 weeks for each firebox. At an average of $1,500 per day plus travel expenses, this amounts to approximately $25,000 for each firebox, or $100,000 for a 4-firebox model line based on crib testing.*
Outside Consultants	$160,000	The average manufacturer will need outside help for design and testing. Testing equipment, knowledge of the test standard, and general guidance is normally offered by outside consultants (not necessarily certified EPA test labs). The average manufacturer will need approximately 300 hours of consulting services per year ($40,000) for 4 years.
Re-tooling	$120,000	For each firebox, new molds and jigs will need to be purchased or produced. Re-tooling charges will reach an estimated $30,000 per firebox, or $120,000 for a 4-firebox model line.
Marketing	$25,000	New pictures will need to be taken and all the current marketing material, including ░░░░░░░░░░░░░░░░ ░░░ ░░░░░░░░░░░░░to be updated.
Total	$1,425,000	Equal to $356,250/model
*Note: The costs originally provided by industry for this table were presumed to be based on crib wood testing, not both crib wood and cord wood testing. Therefore we increased the industry-░░░░░"░░░░░░░"░░░░░░░░ ░░░░░░░ $45,000 shown above) as well as the industry-░░░░░"░░░░░░░░░░░"░░░░░░░░ ░░░░░░░░░░░░ ░░░░░░░░ in order to estimate the additional cost to test with both crib wood and cord wood.		

For this analysis, we used the costs provided in the Table A-1 example, scaled to a single model and spread over a 6-year model development timeframe. We prepared an annualized R&D cost estimate by separating cost elements into direct annual costs (salaries) vs. indirect annual costs (laboratory equipment, retooling and other capital costs). We estimated annual capital costs during the amortized R&D cost period as

---

[107] ░░░░░░░░░ ░░ ░░░░░░░░░░ ░░░░░ ░░ ░░░░░░░░ ░░░░░░░░'░░░░░░░░ement. Prepared by Stove Builder International and United States Stove Company. June 2010.

the fraction that the indirect costs (IAC) are of the Total Annual Cost, approximately 34% annually. Ongoing costs such as taxes, overhead, and other routine expenses would be incurred regardless of the NSPS standard, and are not included in this analysis. Table A-2 shows the estimated annualized cost of $63,850 per model, assuming an amortization period of 6 years and an interest rate of 7%.

Table A-2. Annual Cost Summary: Development of 4 Model Fireboxes[1,2]

Direct Annual Costs (DAC)		
Operator labor	$141,667	Annual salary cost from Table 1, spread over 6 years.
Outside Consultants	$26,667	Annual outside consultant cost from Table 1, spread over 6 years.
Total Direct Costs (DC)	$168,333	
Indirect Annual Costs (IAC)		
Laboratory Equipment[1]	$10,490	The laboratory equipment cost of $50,000 was amortized over 6 years at an interest rate of 7%.
Re-tooling[1]	$25,175	The re-tooling cost of $120,000 was amortized over 6 years at an interest rate of 7%.
Other Capital Costs[1,2]	$51,400	Other capital costs include costs for prototypes ($25,000), test fuel ($30,000+$15,000), testing services ($100,000+$50,000), and marketing ($25,000) and were amortized over 6 years at an interest rate of 7%.
Total Indirect Costs (IAC)	$87,065	
Total Annual Cost	$255,399	Annual cost for development of 4 model fireboxes.
Total Annual Cost	$63,850	Average annual cost per firebox model.

[1] An amortization period of 6 years for laboratory equipment, retooling and other capital costs was chosen based on industry's estimate that approximately 5 to 6 years of R&D are required to bring a product to market.

[2] To estimate the additional cost to test with both cord wood and crib wood, the test fuel industry estimate of $30,000 based on crib only was increased by $15,000 and the testing services industry estimate of $100,000 based on crib only (which covered not only emissions testing but also efficiency and safety testing) was increased by $50,000.

B. Masonry Heaters

Masonry heaters manufacturing cost impacts vary by the type of producer and the type of certification method. According to one manufacturer[108], the masonry heater industry in the U.S. is dominated by the Finnish firm Tulikivi, which manufactures and imports about half of the U.S. masonry heater units installed yearly through its network of installing distributors. The same manufacturer said that the second largest producer is a

---

[108] Comments: Residential Solid Fuel Burning Appliance SBREFA Process. Product Category: Masonry Heaters. July 13, 2010. Timothy Seaton, Timely Construction, Inc. p. 5.

Canadian firm, Temp-□□□□□□□□□□□□ □□□ □□□□□□□□□□ □□□□□□□□□□□□"□□□□□"□□□□□ □□□□□□□□□□□□□□□□□ □□□□□□□□□□□□□□□□□□□ □□□ than 100 (or at least fewer than 200) generating any masonry revenue at all. Some commercial operations sell core units and/or design kits based on their own design, and other sell units they license from other U.S. or foreign companies. Finally, some units are custom built. Based on this information, we assumed that 50 percent of masonry heaters sold per year in the U.S. are Tulikivi models and 35 percent are sold by other large manufacturers. The remaining 15 percent of units are sold by independent small volume contractors.

There are three major cost components to consider in evaluating the potential cost impacts of the proposed NSPS: research and development (R&D), certification testing, and licensing fees for use of a computer software package approved for use in certifying a model design. According to information provided by one manufacturer[109], R&D costs for a masonry heater may be estimated at $250,000 (compared to the $340,000 for other wood heater appliance models noted in Table A-1). We assumed R&D costs were the same as for other wood heater appliances □that is, $63,850 annually for a 6-year R&D amortization period. For facilities conducting R&D, these costs include the costs for certification testing. We estimate that the cost of testing a heater design in an EPA accredited lab to be approximately $10,000.[110]

This cost analysis also makes use of a unique software package based on a European masonry heater design standard. This standard has been verified in the laboratory and under field conditions to produce masonry heaters that would meet the proposed NSPS emission limits. The software produces for printout a certification for a given design application and the design definition documents as well as operating instructions customized to the given design, so that the software verification and certification record is created for and attached to the design. The resulting documents can be submitted as part of the certification application. The cost of this software to the user is approximately 1,000 Euros (approximately $1,500) for the package with a 300 Euro (approximately $450) annual fee that commences in the second year following purchase.[111]

II.   Estimated Manufacturer Cost Impacts

A. NSPS Options

We developed two primary options to evaluate the cost impacts of implementing new or revised NSPS standards for the residential wood heating appliance industry □namely, the propos□□□□□□□□□□"□□□□□□□"□□□□□

---

[109] Comments: Residential Solid Fuel Burning Appliance SBREFA Process. Product Category: Masonry Heaters. July 13, 2010. Timothy Seaton, Timely Construction, Inc. p. 14.

[110] Letter to Lucinda Power, EPA, from Brian Klipfel, Fire Works Masonry. September 10, 2010.

[111] E-mail from Timothy Seaton, Timely Construction Company, to Gil Wood, USEPA. April 21, 2011.

□□□□□□□□□□□□□□□□□□"□□□□□□□□□"□□□□ □□□□□□□□□□□□□□□□□□□□□□□□□□ □□ □□□□□□[112], the cost analyses assume specified phase-in schedules which vary between the Proposal and Alternative.

The Proposal represents a scenario where all types of appliances (except masonry heaters) would be required to meet a specified Step 1 emission limit upon promulgation in 2014 and then a stricter Step 2 emission limit five years later in 2019. Under this Proposal, large manufacturers of masonry heaters would also be required to meet a specified Step 1 emission limit upon promulgation in 2014, while small custom manufacturers would not be required to meet this Step 1 emission limit until 2019. There is no Step 2 emission limit for masonry heaters being proposed. The Alternative approach represents a scenario where all types of appliances (except masonry heaters) would be required to meet a specified Step 1 emission limit upon promulgation in 2014 and then an interim Alternative Step 2 emission limit three years after promulgation in 2017, followed by a stricter Alternative Step 3 emission limit eight years after promulgation in 2022. Masonry heaters would be regulated the same way under this Alternative approach as under the Proposal, as explained above. Note that the Proposal Step 2 emission limit is the same as the Alternative Step 3 emission limit, but the compliance dates differ for this stricter limit (i.e., 2019 versus 2022). These implementation periods represent a tentative schedule and are subject to change.

Following is a summary of the NSPS implementation assumptions for each appliance type, grouped according to NSPS Subparts. This summary describes the specific emission limits under the Proposal and the Alternative approach.

Subpart AAA "□□□ □□□□□" □□□
1. Adjustable burn rate, single burn rate, and pellet stoves: **Proposal**: Step 1 limit of 4.5 g/hr upon promulgation in 2014; and Step 2 limit of 1.3 g/hr five years after promulgation in 2019. **Alternative**: Step 1 limit of 4.5 g/hr upon promulgation in 2014; Step 2 limit of 2.5 g/hr three years after promulgation in 2017; and Step 3 limit of 1.3 g/hr eight years after promulgation in 2022.

   Notes on appliances already meeting these limits: The Step 1 limit is the 1995 Washington State standard for non-catalytic stoves; the Alternative Step 2 limit is the 1995 Washington State standard for catalytic stoves; and the proposed Step 2 (Alternative Step 3) limit is already met by the top performing catalytic, non-catalytic and pellet stove models, according to industry data.[113] Although previously unregulated and a less developed technology than adjustable burn rate stoves, single burn rate stove designs have been undergoing R&D in anticipation of the proposed NSPS and cleaner designs are nearly market-ready.[114]
   Under both the Proposal and the Alternative, adjustable burn rate wood stoves (both catalytic and non-catalytic), pellet stoves, and single burn rate stoves would face a Step 1 standard that is based on the current Washington state standards of 4.5 g/hr emissions for non-catalytic stoves. Industry data[10]

---

[112] Memo to Gil Wood, USEPA, from Jill Mozier, EC/R, Inc. Unit Cost Estimates of Residential Wood Heating Appliances. February 21, 2013.

[113] Letter to Gil Wood, EPA, from Robert Ferguson, Ferguson, Andors, & Company. April 30, 2010. EC/R made minor changes to □□□□□□□□□□□□□□□□□□□□□□□□ □□ □□□□ □□□□□ □□□□□□□□□□□□□□□ □□□□□□□□□□□□□□"□□□□□□□□ Heater Dat□□□□" □□□□ □□□□□□□ □□□□□□

[114] 2/8/13 telephone discussion between Gil Wood, USEPA, and a manufacturer of single burn rate stoves.

from 2010 indicate that 90% (130 out of 145 catalytic, non-catalytic and pellet stoves combined) already meet this limit, as shown in Table A-1 of Attachment A ☐☐☐☐☐☐ ☐☐☐☐☐☐☐☐☐' ☐☐☐☐memo. The tables in that attachment show the emissions performance of wood stoves that are currently certified for which we have reproducible emissions data. This list was initially prepared by the Hearth, Patio and Barbecue Association (HPBA), and includes the results of an effort to delete models that are no longer manufactured and remove duplicate certifications.[10] Communication from HPBA confirmed that there were 110 non-catalytic wood stoves and 15 catalytic wood stoves being manufactured in 2010 by 34 manufacturers or importers of EPA certified wood heaters.[115] We believe that manufacturers will focus on existing models that meet the Washington State limits in order to comply with the Step 1 standard.

Furthermore, the HPBA data[10] indicates that 35% (52 out of 145 catalytic, non-catalytic and pellet stoves combined) already meet the interim Alternative Step 2 limit of 2.5 g/hr. Finally, this same dataset indicates that 10% (14 out of 145 catalytic, non-catalytic and pellet stoves combined) already meet the strictest proposed Step 2 standard (which is also the Alternative Step 3 standard). Stoves meeting these limits are shown in Table A-1 of Attachment A, in which green rows indicate compliance with proposed Step 2/Alternative Step 3, green and blue rows indicate compliance with interim Alternative Step 2, and green, blue, and orange rows indicate compliance with Step 1. Attachment A also includes tables showing this same color-coded compliance for the individual catalytic (Table A-2) and pellet stove (Table A-3) datasets.

Based on this emission data for adjustable burn rate and pellet stoves, for our cost analysis we assumed that while major R&D efforts are not needed to meet Step 1, R&D efforts would be required to meet the strictest standard.

Single burn rate stoves would also be subject to these stepped standards under Subpart AAA. Given the relative lack of previous regulation of these appliance types in the United States, we believe that manufacturers will have to redesign all such models to meet the proposed emission limits, and reflected that major re-design effort in our cost estimates.

Our specific cost and implementation assumptions under both the Proposal and Alternative scenarios are noted below in Sections B & D.

Subpart QQQQ ☐"☐☐☐☐☐☐☐☐☐☐"☐☐

2. Hydronic heaters (both outdoor and indoor): **Proposal**: Step 1 limit of 0.32 lb/mm BTU heat output upon promulgation in 2014; and Step 2 limit of 0.06 lb/mm BTU heat output five years after promulgation in 2019. **Alternative**: Step 1 limit of 0.32 lb/mm BTU heat output upon promulgation in 2014; Step 2 limit of 0.15 lb/mm BTU heat output three years after promulgation in 2017; and Step 3 limit of 0.06 lb/mm BTU heat output eight years after promulgation in 2022.

Notes on appliances already meeting these limits☐☐☐☐☐☐☐☐☐☐☐ ☐☐☐☐☐☐☐☐☐☐"☐☐☐☐☐"☐ voluntary program limit already met by 36 hydronic heater models (27 cord wood and 9 pellet models) built by 17 U.S. manufacturers; the Alternative Step 2 limit is already met by 11 hydronic heater models (6 cord wood and 5 pellet models) built by 6 U.S. manufacturers; and the proposed Step 2 (Alternative Step 3) limit is already met by 4 hydronic heater models (2 cord wood and 2 pellet models) built by 2 U.S. manufacturers[116], as well as over 100 European manufacturers per test method EN 303-05.[117]

---

[115]Letter to Lucinda Power, EPA, from John Crouch, HPBA. September 12, 2010.

[116] See list of cleaner hydronic heaters participa☐☐☐☐☐☐☐☐☐' ☐☐☐☐☐☐☐☐☐☐☐☐ ☐☐http://www.epa.gov/burnwise/owhhlist.html

[117] European Wood-Heating Technology Survey: An Overview of Combustion Principles and the Energy and Emissions Performance Characteristics of Commercially Available Systems in Austria, Germany, Denmark, Norway, and Sweden; Final Report; Prepared for the New York State Energy Research and Development Authority; NYSERDA Report 10☐01; April 2010.

understand that it is dominated by a few manufacturers in terms of the bulk of sales, and these manufacturers have qualifying units at some or all of the stepped emission limits, as noted above. Therefore, on a sales-weighted basis, only a percentage of the hydronic heater models currently sold would be required to undertake R&D to meet Step 1, with higher percentages needing R&D to meet the interim Alternative Step 2 and the proposed Step 2/Alternative Step 3 limits. However, we assumed that all hydronic heater models would begin R&D in 2013 to meet the stricter phased-in standards. Our specific cost and implementation assumptions under both the Proposal and Alternative scenarios are noted below in Sections B & D.

3. <u>Forced Air Furnaces</u>: **Proposal**: Step 1 limit of 0.93 lb/mm BTU heat output upon promulgation in 2014; and Step 2 limit of 0.06 lb/mm BTU heat output five years after promulgation in 2019. **Alternative**: Step 1 limit of 0.93 lb/mm BTU heat output upon promulgation in 2014; Step 2 limit of 0.15 lb/mm BTU heat output three years after promulgation in 2017; and Step 3 limit of 0.06 lb/mm BTU heat output eight years after promulgation in 2022.

4. **Notes on appliances already meeting these limits**: The Step 1 limit is based on test data from development of Canadian standard B415.1-10[118] and conversation with industry regarding cleaner forced air furnace models currently being tested in R&D[119]. Forced air furnace designs able to meet the Alternative Step 2 and proposed Step 2 (Alternative Step 3) limits may be based on technology transferred from hydronic heater designs. Given the relative lack of previous regulation of these appliance types in the United States, we assume in our cost analysis that manufacturers will have to redesign all such models to meet the proposed emission limits, and reflect that major re-design effort in our cost estimates. Our specific cost and implementation assumptions under both the Proposal and Alternative scenarios are noted below in Sections B & D.

<u>Subpart RRRR</u> (masonry heaters):

5. <u>Masonry Heaters</u>: **Proposal / Alternative** (same): Step 1 limit of 0.32 lb/mm BTU heat output upon promulgation in 2014 for large manufacturers (defined as manufacturers constructing ≥ 15 masonry heaters per year), with a 5-year (2019) small volume manufacturer compliance extension (for companies constructing < 15 units/year). No other phased-in limits are being proposed.

Notes on appliances already meeting these limits: Based on data submitted by the Masonry Heater Association[120], over 10 models already meet this limit. The masonry heater market is dominated by a few large manufacturers and many small custom manufacturers, and these segments of the market will take different approaches to come into compliance based on their models (i.e., a couple models will undergo R&D, while others who already meet the standards will certify through a conventional test or through a software product). Our cost analysis segmented the market accordingly and based our cost estimates on specific assumptions relevant to each segment of the market. These specific cost and implementation assumptions (which do not differ for the Proposal and Alternative) are explained below in Sections C & D.

B.   Estimated Manufacturer Costs – General Approach

---

[118] CSA B415.1-10, Performance Testing of Solid-Fuel-Burning Heating Appliances. Appendix D. March 2010.
[119] 2/8/13 telephone discussion between Gil Wood, USEPA, and a manufacturer of forced air furnaces.
[120] Attachment to 3/25/2011 e-mail from Timothy Seaton of Timely Construction to Gil Wood and Mike Toney of USEPA

Manufacturers have told us that is takes several years to develop new models. We have spread the annualized R&D costs (shown in Table A-2) over 6 years to represent the time and funds needed to develop the complying models. For the purposes of our cost estimate, we have assumed that when the NSPS revisions are proposed, all manufacturers will begin serious efforts to develop complying models, although for many manufacturers we also know that they have been involved in intensive R&D efforts in anticipation of the proposed rule.

We estimated both the average annual cost to manufacturers of each appliance type and then extended those costs to nationwide total annual costs. The basic components to ▯▯▯▯▯ ▯▯▯▯▯▯▯▯▯'▯▯▯▯▯ ▯▯▯▯▯▯▯▯▯▯▯▯ are:

6. Annualized R&D cost;
7. Ongoing annual Certification cost; and
8. Ongoing annual Reporting and Record Keeping cost.

The Annualized R&D costs (shown in Table A-2, and based on the Table A-1 costs) are by far the largest cost component and we have applied these costs to most models in our cost analysis ▯especially to models in previously unregulated appliance categories ▯in order to present a reasonable estimate of the costs. For example, as noted above, instead of estimating the number of hydronic heater models that already meet a specific limit and will therefore merely need to certify their emissions rather than undergo R&D, we instead assumed that 100% of hydronic heater models will undergo R&D beginning in 2013. We made similar assumptions for single burn rate stoves and forced air furnaces.

Under the Proposal scenario, one round of R&D is assumed ▯beginning in 2013 and ending in 2018 ▯in order to meet the proposed Step 2 limit. Under the Alternative scenario, two rounds of R&D are assumed for all appliances except masonry heaters (for which there is only one standard with no additional phased-in standards to meet). Under the Alternative scenario, the first R&D round begins in 2013 and the second round begins in 2017 (which causes overlapping R&D costs in years 2017 and 2018 in this analysis) ▯in order to meet the interim Alternative Step 2 limit in 2017 and the Alternative Step 3 limit in 2022. We also assumed that of the models undergoing the first round of R&D costs, 80% of these models undergo the second round of R&D costs in the Alternative scenario (i.e., we assumed that only 20% of models achieve the strictest limit in the first round of R&D).

Furthermore, for appliances like single burn rate stoves and forced air furnaces, which were previously unregulated (and also were not pushed technologically by a voluntary program, as hydronic heaters were), we have conservatively doubled R&D costs during years 2013 and 2014. This doubling of R&D cost estimates is to represent an intensification of the R&D efforts to meet the Step 1 limit and begin development of models to meet the stricter stepped limits ☐R&D efforts which industry has indicated are already ongoing.[121]

Note that all manufacturers, except for wood stoves that are subject to the current 1988 NSPS, will face ongoing certification costs above baseline conditions. However, in the 2013 to 2018 timeframe under the Proposal scenario and in the 2013 to 2022 timeframe under the Alternative scenario, we have incorporated these costs as part of the overall R&D expenditures. After 2018 under the Proposal scenario and after 2022 under the Alternative scenario, the ongoing certification costs will be the only NSPS related costs faced by manufacturers besides ongoing reporting and recordkeeping costs.

Regarding certification costs, we have assumed a cost of $10,000 per model for pellet stoves, single burn rate stoves and masonry heaters; and we have assumed a cost of $20,000 per model for hydronic heaters and forced air furnaces.[122] We have spread these costs out over the 5 year certification life, assuming annual certification costs for one-fifth of the models.

For example, pellet stoves will incur certification costs in advance of complying with more stringent limits. As explained in Section A and shown in Table A-3 of Attachment A ☐☐☐☐☐☐ ☐☐☐☐☐☐☐☐☐'☐☐☐☐☐☐ ☐☐ ☐, approximately 30 percent of existing pellet stove models are expected to comply with the proposed Step 2 and Alternative Step 3 standard. However, in order to be sold, these stove models would now be required to demonstrate compliance with an emissions limit, incurring an upfront cost of $10,000 per model to become certified. We have assumed that one fifth of the pellet stove models will certify in any given year.

We based reporting and recordkeeping (R&R) costs on the annual average costs derived from development of the Information Collection Request (ICR) supporting statements[123]. These are annual estimates of the ongoing R&R burden to manufacturers associated with the Proposal and Alternative scenarios. (We do not expect the R&R burden to differ substantially between the two scenarios.)

The certification and reporting and recordkeeping costs were estimated to be incurred by manufacturers for the full 20-year model design lifespan.[124] Under the Proposal, we estimated costs from 2013 through 2038 ☐ that is, 20 years after the 2019 compliance year marking the beginning of the model lifespan designed to meet the Proposal Step 2 limit. Under the Alternative, we estimated costs from 2013 through 2041 ☐that is, 20 years after the 2022 compliance year marking the beginning of the model lifespan designed to meet the Alternative Step 3 limit.

---

[121] 2/8/13 telephone discussion between Gil Wood, USEPA, and a manufacturer of forced air furnaces and single burn rate stoves.
[122] Conversation with Dennis Brazier, Central Boiler. August 9, 2010.
[123] ICR Supporting Statements for the Proposed NSPS Subparts have not been finalized as of the date of this memo.
[124] Memo to Gil Wood, USEPA, from Jill Mozier, EC/R, Inc. Unit Cost Estimates of Residential Wood Heating Appliances. February 21, 2013.

C.   Estimated Manufacturer Costs – Masonry Heaters

As noted above, we addressed masonry heaters in a way which segmented the costs in keeping with the masonry heater market. There are three scenarios for potential cost impacts for large masonry heater manufacturers. In the case of Tulikivi and some U.S. firms, e.g., Timely Construction, these companies have already invested in R&D in order to gain access to U.S. markets which restrict sales (e.g., Colorado) of uncertified units. These companies will face testing costs only, with an assumed total of nine tests conducted prior to the proposed compliance date (i.e., to certify a total of nine model lines). For purposes of our cost analysis, we assumed that two additional companies will conduct R&D to develop two new models each to meet the proposed NSPS. Finally, we have been told that Tulikivi will use the software certification approach to certify up to eight additional models. We also project that the remaining 15 percent of custom built units will use the software certification approach to certify compliance with the proposed NSPS starting in 2013 (estimated date of the proposed standards) and that they will continue to renew their license in the following years.

As explained in the unit cost memo[21], we used data in the Frost & Sullivan Market (F&S) report[125] on 2008 masonry heater shipments by product category and F&S revenue forecasts which incorporated the weak economy in years 2009 and 2010, to calculate the reduced number of shipments in years 2009 and 2010. For years 2011 through 2038 (for the Proposal) and 2011 through 2041 (for the Alternative) estimated shipments are based on a forecasted revenue growth rate of 2.0%, in keeping with the average annual growth in real GDP predicted by the US Bureau of Economic Analysis.[126] For masonry heaters, our estimate of the number of custom built models is based on the average number of models sold per year in the 15 percent model category (i.e., 85 per year). We assumed each custom manufacturer would sell 2 models per year, for a total of 42 manufacturers participating in the software certification option.

Under both the Proposal and Alternative scenarios, most sales-weighted masonry heater units face a 2014 Step 1 compliance date with no other phased-in limits. However, under both the Proposal and Alternative scenarios, companies that sell fewer than 15 units per year have until 2019 to come into compliance. We have assumed that the large manufacturers will comply by 2014 for the units that only require testing and/or software certification, with those expenditures incurred annually starting in 2013. We also assumed that the 15 percent of custom built units will comply by 2019, but will begin certifying their units using the software certification approach as early as 2013, as noted above, as a selling point for their services. For those companies that start R&D when the NSPS is proposed in 2013, we have assumed that they will spread these costs over the 6-year period from 2013 through 2018 for the four models affected, under both the Proposal and Alternative scenarios.

---

[125] Market Research and Report on North American Residential Wood Heaters, Fireplaces, and Hearth Heating Products Market. Prepared by Frost & Sullivan. April 26, 2010. P. 31-32.

[126] 2013 Global Outlook projections prepared by the Conference Board in November 2012; http://www.conference-board.org/data/globaloutlook.cfm

D.   Estimated Manufacturer Costs – Specific Assumptions & Resulting Costs

Table A-3a shows the estimated annual cost per manufacturer under the Proposal for all appliances. Table A-3b shows the nationwide annual costs under the Proposal. The footnotes associated with the tables (not included in the tables shown on the following pages) help better explain the details we assumed for the cost analysis and are listed below. For the Proposal, the footnoted assumptions underlying Tables A-3a and A-3b are:

1. Nationwide Annual Cost assumes R&D investment is amortized over 6 years (2013 through 2018). Ongoing certification costs are incurred through 2038 (based on a model brought to market in 2019 with a lifespan of 20 years), except for woodstoves which already incur certification costs under the existing NSPS.

2. Since certification is required every 5 years (except for the software certification option for masonry heaters), it is assumed that certification costs will be spread out so that 1/5 of the models certify each year.

3. This analysis considers additional costs resulting from the proposed NSPS. For wood stoves, the analysis assumes that 5% meet Step 2 already so that 95% of the models will undergo re-design to meet the Step 2 level. The costs modeled for years 2020 through 2038 exclude the ongoing certification costs and ongoing reporting and recordkeeping costs incurred by wood stove manufacturers who already had to certify and report under the existing NSPS.

4. For pellet stoves, the analysis assumes that 30% meet Step 2 already so that 70% of models undergo R&D re-design to meet Step 2. The R&D budget includes certification costs. The analysis also assumes that the 30% of the pellet stove models which already meet Step 2 will certify in an ongoing basis starting in 2013. The analysis reflects the certification costs beginning in 2013 for the 30% of models meeting Step 2, and beginning in 2019 for the remaining 70% of models which underwent R&D re-design.

5. Based on conversations with industry (2/2013), single burn rate stoves and forced air furnaces have been undergoing R&D prior to 2013 to develop cleaner models. Because these devices were previously unregulated and may need to transfer technology from adjustable burn rate stoves and hydronic heaters respectively, this analysis assumes that these efforts will intensify in 2013 and 2014. Therefore estimated R&D costs are doubled in 2013 and 2014 in order to meet the 2014 Step 1 standard while also beginning R&D to develop models to meet the more stringent 2019 Step 2 standard.

6. For single burn rate stoves, forced air furnaces, and hydronic heating systems, the analysis assumes that only a small percentage meet Step 2 so that approximately 100% of the models undergo R&D re-design to meet Step 2. The R&D budget includes certification costs. Ongoing certification costs for the re-designed models are reflected in this analysis beginning in 2019.

7. For masonry heaters, the cost analysis assumes one round of R&D to meet 0.32 lb/mmBTU standard (no additional stepped standards, although large manufacturers will be required to meet the limit in 2014, while small volume manufacturers will be given a 5 year extension until 2019 to meet the limit). For masonry heater manufacturers using software certification, the analysis assumes the purchased software will be used for certifying all models developed by that manufacturer.

8. Reporting and recordkeeping costs (R&R) [for all appliances but masonry heaters] are based on the annual average costs derived from the ICR and are estimates of the ongoing R&R burden to manufacturers associated with the proposed NSPS. The annual average nationwide R&R burden estimated to manufacturers for Subpart AAA is $440,443, and for Subpart QQQQ is $119,249. These R&R costs do not include the R&R burden to laboratories; the annual average nationwide R&R burden incurred by laboratories subject to requirements under Subpart AAA is estimated to be $75,745, and incurred by laboratories subject to requirements under Subpart QQQQ is estimated to be $50,496.

9. [Masonry Heater] Reporting and recordkeeping costs (R&R) are based on the annual average costs derived from the ICR and are estimates of the ongoing R&R burden to manufacturers associated with the proposed NSPS. The annual average nationwide R&R burden estimated to manufacturers for Subpart RRRR is $98,788 for small/custom masonry heater manufacturers and $25,929 for large masonry heater manufacturers. These R&R costs do not include the R&R burden to laboratories; the annual average nationwide R&R burden incurred by laboratories subject to requirements under Subpart RRRR is estimated to be $37,872.

Table A-4a shows the estimated annual cost per manufacturer under the Alternative approach for all appliances. Table A-4b shows the nationwide annual costs under the Alternative. For the Alternative approach, the footnoted assumptions underlying Tables 4a and 4b (where different from the footnotes listed above) are:

1. Nationwide Annual Cost assumes R&D investment is amortized over 6 years (round one from 2013 through 2018 and round two from 2017 through 2022). Ongoing certification costs are incurred through 2041 (based on a model brought to market in 2022 with a lifespan of 20 years), except for woodstoves which already incur certification costs under the existing NSPS.

2. (Same as above)

3. This analysis considers additional costs resulting from the proposed NSPS. For wood stoves, the analysis assumes that 5% meet Step 3 already so that 95% of the models will undergo re-design in round one, and 80% of those 95% will require another round of R&D to meet the Step 3 level. The costs exclude the ongoing certification costs and ongoing reporting and recordkeeping costs incurred by wood stove manufacturers who already had to certify and report under the existing NSPS.

4. For pellet stoves, the analysis assumes that 30% meet Step 3 already so that 70% of models undergo re-design in round one, and 80% of those 70% require another round of R&D to meet Step 3. The R&D budget includes certification costs. The analysis also assumes that the 30% of the pellet stove models which already meet Step 3 will certify in an ongoing basis starting in 2013.

5. Based on conversations with industry (2/2013), single burn rate stoves and forced air furnaces have been undergoing R&D prior to 2013 to develop cleaner models. Because these devices were previously unregulated and may need to transfer technology from adjustable burn rate stoves and hydronic heaters respectively, this analysis assumes that these efforts will intensify in 2013 and 2014. Therefore estimated R&D costs are doubled in 2013 and 2014 in order to meet the 2014 Step 1 standard while also beginning R&D to develop models to meet the more stringent 2017 Step 2 and 2022 Step 3 standards.

6. For single burn rate stoves, forced air furnaces, and hydronic heating systems, the analysis assumes that only a small percentage meet Step 3 so that approximately 100% of the models undergo re-design in round one, and 80% require another round of R&D to meet Step 3. The R&D budget includes certification costs.
7. (Same as above)
8. (Same as above)
9. (Same as above)

Table A-3a. Average Annual Cost per Manufacturer under the Proposal

NSPS Subpart	Appliance Type	# Manufacturers	# Models	Average Annual Cost per Manufacturer based on 6-year R&D round (2013-2018) to meet Step 1 and Step 2 limits with ongoing certification costs (through 2038)							
				2013	(Step 1 compliance) 2014	2015	2016	2017	2018	(Step 2 compliance) 2019	2020 through 2038[2]
	Wood Stoves (R&D)[3]	34	125	$223,004	$223,004	$223,004	$223,004	$223,004	$223,004	$0	$0
	Pellet Stoves (R&D, R&R)[4,8]	29	125	$199,681	$199,681	$199,681	$199,681	$199,681	$199,681	$7,031	$7,031
	Pellet Stoves (certification)[4]	29	125	$2,586	$2,586	$2,586	$2,586	$2,586	$2,586	$8,621	$8,621
	Single Burn Rate Stoves (R&D, R&R, cert.)[5,6,8]	3	20	$862,204	$862,204	$436,539	$436,539	$436,539	$436,539	$24,208	$24,208
AAA: Room Heaters											
	Forced Air Furnaces (R&D, R&R, cert.)[5,6,8]	7	50	$917,148	$917,148	$461,079	$461,079	$461,079	$461,079	$33,582	$33,582
	Hydronic Heating Systems (R&D, R&R, cert.)[6,8]	30	120	$258,204	$258,204	$258,204	$258,204	$258,204	$258,204	$18,806	$18,806
QQQQ: Central Heaters											
	MH - large companies (R&D, R&R, cert.)[9]	2	4	$130,169	$130,169	$130,169	$130,169	$130,169	$130,169	$6,469	$6,469
	MH - large companies (R&R, cert.)[9]	3	9	$9,704	$9,704	$9,704	$9,704	$9,704	$9,704	$9,704	$9,704
	MH - large companies (R&R, software cert.)[7,9]	1	8	$11,378	$10,328	$10,328	$10,328	$10,328	$10,328	$10,328	$10,328
	MH - small companies (R&R, software cert.)[7,9]	42	85	$3,852	$2,802	$2,802	$2,802	$2,802	$2,802	$2,802	$2,802
RRRR: Masonry Heaters											

Table A-3b. Nationwide Annual Costs under the Proposal

NSPS Subpart	Appliance Type	# Manufac-turers	# Models	Nationwide Annual Costs[1]							
				2013	(Step 1 compliance) 2014	2015	2016	2017	2018	(Step 2 compliance) 2019	2020 through 2038[2]
	Wood Stoves (R&D)[3]	34	125	$7,582,146	$7,582,146	$7,582,146	$7,582,146	$7,582,146	$7,582,146	$0	$0
	Pellet Stoves (R&D, R&R)[4,8]	29	125	$5,790,753	$5,790,753	$5,790,753	$5,790,753	$5,790,753	$5,790,753	$203,909	$203,909
	Pellet Stoves (certification)[4]	29	125	$75,000	$75,000	$75,000	$75,000	$75,000	$75,000	$250,000	$250,000
	Single Burn Rate Stoves (R&D, R&R, cert.)[5,6,8]	3	20	$2,586,611	$2,586,611	$1,309,618	$1,309,618	$1,309,618	$1,309,618	$72,625	$72,625
AAA: Room Heaters				$16,034,510	$16,034,510	$14,757,517	$14,757,517	$14,757,517	$14,757,517	$526,534	$526,534
	Forced Air Furnaces (R&D, R&R, cert.)[5,6,8]	7	50	$6,420,038	$6,420,038	$3,227,556	$3,227,556	$3,227,556	$3,227,556	$235,073	$235,073
	Hydronic Heating Systems (R&D, R&R, cert.)[6,8]	30	120	$7,746,133	$7,746,133	$7,746,133	$7,746,133	$7,746,133	$7,746,133	$564,176	$564,176
QQQQ: Central Heaters				$14,166,171	$14,166,171	$10,973,689	$10,973,689	$10,973,689	$10,973,689	$799,249	$799,249
	MH - large companies (R&D, R&R, cert.)[9]	2	4	$260,337	$260,337	$260,337	$260,337	$260,337	$260,337	$12,939	$12,939
	MH - large companies (R&R, cert.)[9]	3	9	$29,112	$29,112	$29,112	$29,112	$29,112	$29,112	$29,112	$29,112
	MH - large companies (R&R, software cert.)[7,9]	1	8	$11,378	$10,328	$10,328	$10,328	$10,328	$10,328	$10,328	$10,328
	MH - small companies (R&R, software cert.)[7,9]	42	85	$161,788	$117,688	$117,688	$117,688	$117,688	$117,688	$117,688	$117,688
RRRR: Masonry Heaters				$462,616	$417,466	$417,466	$417,466	$417,466	$417,466	$170,067	$170,067
Annual Cost of the Rule				$30,663,297	$30,618,147	$26,148,672	$26,148,672	$26,148,672	$26,148,672	$1,495,850	$1,495,850

Table A-4a. Average Annual Cost per Manufacturer under the Alternative Approach

NSPS Subpart	Appliance Type	# Manufacturers	# Models	Average Annual Cost per Manufacturer based on initial 6-year R&D round (2013-2018) and second R&D round (2017-2022) to meet Step 3 level with ongoing certification costs (through 2041)										
				2013	(Step 1 compliance) 2014	2015	2016	(Step 2 compliance) 2017	2018	2019	2020	2021	(Step 3 compliance) 2022	2023 through 2041 [2]
AAA: Room Heaters														
	Wood Stoves (R&D)[3]	34	125	$223,004	$223,004	$223,004	$223,004	$394,080	$394,080	$171,076	$171,076	$171,076	$171,076	$0
	Pellet Stoves (R&D, R&R)[4,8]	29	125	$199,681	$199,681	$199,681	$199,681	$347,471	$347,471	$154,821	$154,821	$154,821	$154,821	$7,031
	Pellet Stoves (certification)[4]	29	125	$2,586	$2,586	$2,586	$2,586	$2,586	$2,586	$3,793	$3,793	$3,793	$3,793	$8,621
	Single Burn Rate Stoves (R&D, R&R, cert.)[5,6,8]	3	20	$862,204	$862,204	$436,539	$436,539	$763,085	$763,085	$340,087	$340,087	$340,087	$340,087	$24,208
QQQQ: Central Heaters														
	Forced Air Furnaces (R&D, R&R, cert.)[5,6,8]	7	50	$917,148	$917,148	$461,079	$461,079	$810,949	$810,949	$360,594	$360,594	$360,594	$360,594	$33,582
	Hydronic Heating Systems (R&D, R&R, cert.)[6,8]	30	120	$258,204	$258,204	$258,204	$258,204	$454,131	$454,131	$201,933	$201,933	$201,933	$201,933	$18,806
	MH – large companies (R&D, R&R, cert.)[9]	2	4	$130,169	$130,169	$130,169	$130,169	$130,169	$130,169	$6,469	$6,469	$6,469	$6,469	$6,469
	MH – large companies (R&R, cert.)[9]	3	9	$9,704	$9,704	$9,704	$9,704	$9,704	$9,704	$9,704	$9,704	$9,704	$9,704	$9,704
	MH – large companies (R&R, software cert.)[7,9]	1	8	$11,378	$10,328	$10,328	$10,328	$10,328	$10,328	$10,328	$10,328	$10,328	$10,328	$10,328
	MH – small companies (R&R, software cert.)[7,9]	42	85	$3,852	$2,802	$2,802	$2,802	$2,802	$2,802	$2,802	$2,802	$2,802	$2,802	$2,802
RRRR: Masonry Heaters														

Table A-4b. Nationwide Annual Costs under the Alternative Approach

NSPS Subpart	Appliance Type	# Manufacturers	# Models	Nationwide Annual Costs[1]										
				2013	2014	2015	2016	2017	2018	2019	2020	2021	2022	2023 through 2041[2]
AAA: Room Heaters	Wood Stoves (R&D)[3]	34	125	$7,582,146	$7,582,146	$7,582,146	$7,582,146	$13,398,730	$13,398,730	$5,816,584	$5,816,584	$5,816,584	$5,816,584	$0
	Pellet Stoves (R&D, R&R)[4,8]	29	125	$5,790,753	$5,790,753	$5,790,753	$5,790,753	$10,076,657	$10,076,657	$4,489,813	$4,489,813	$4,489,813	$4,489,813	$203,909
	Pellet Stoves (certification)[4]	29	125	$75,000	$75,000	$75,000	$75,000	$75,000	$75,000	$110,000	$110,000	$110,000	$110,000	$250,000
	Single Burn Rate Stoves (R&D, R&R, cert.)[5,6,8]	3	20	$2,586,611	$2,586,611	$1,309,618	$1,309,618	$2,289,254	$2,289,254	$1,020,261	$1,020,261	$1,020,261	$1,020,261	$72,625
				$16,034,510	$16,034,510	$14,757,517	$14,757,517	$25,839,640	$25,839,640	$11,436,657	$11,436,657	$11,436,657	$11,436,657	$526,534
QQQQ: Central Heaters	Forced Air Furnaces (R&D, R&R, cert.)[5,6,8]	7	50	$6,420,038	$6,420,038	$3,227,556	$3,227,556	$5,676,644	$5,676,644	$2,524,161	$2,524,161	$2,524,161	$2,524,161	$235,073
	Hydronic Heating Systems (R&D, R&R, cert.)[6,8]	30	120	$7,746,133	$7,746,133	$7,746,133	$7,746,133	$13,623,945	$13,623,945	$6,057,987	$6,057,987	$6,057,987	$6,057,987	$564,176
				$14,166,171	$14,166,171	$10,973,689	$10,973,689	$19,300,588	$19,300,588	$8,582,148	$8,582,148	$8,582,148	$8,582,148	$799,249
	MH - large companies (R&D, R&R, cert.)[9]	2	4	$260,337	$260,337	$260,337	$260,337	$260,337	$260,337	$12,939	$12,939	$12,939	$12,939	$12,939
	MH - large companies (R&R, cert.)[9]	3	9	$29,112	$29,112	$29,112	$29,112	$29,112	$29,112	$29,112	$29,112	$29,112	$29,112	$29,112
	MH - large companies (R&R, software cert.)[7,9]	1	8	$11,378	$10,328	$10,328	$10,328	$10,328	$10,328	$10,328	$10,328	$10,328	$10,328	$10,328
	MH - small companies (R&R, software cert.)[7,9]	42	85	$161,788	$117,688	$117,688	$117,688	$117,688	$117,688	$117,688	$117,688	$117,688	$117,688	$117,688
RRRR: Masonry Heaters				$462,616	$417,466	$417,466	$417,466	$417,466	$417,466	$170,067	$170,067	$170,067	$170,067	$170,067
Annual Cost of the Rule				$30,663,297	$30,618,147	$26,148,672	$26,148,672	$45,557,694	$45,557,694	$20,188,873	$20,188,873	$20,188,873	$20,188,873	$1,495,850

It should be noted that Tables A-3a, A-3b, A-4a, and A-4b are based on a 7% interest rate and are in 2010 dollars($). We also prepared these cost estimates based on a 3% interest rate. Note also that costs vary by appliance type based on the average number of models per manufacturer. The estimate of the number of model types are described in the unit cost memo.[127] For numbers of manufacturers, we started with HPBA data, modified based on internet searches of manufacturers of the major appliance types.[128]

The total nationwide cost of the rule from years 2014 through 2022 for the Proposal and the Alternative differ based on the underlying cost and implementation assumptions described in this memo, and are summarized below in Table A-5.

Table A-5. Nationwide Annual Cost of the Rule under the Proposal and Alternative Approach

Year	Cost under Proposal (2010$)	Cost under Alternative Approach (2010$)
2014	30,618,147	30,618,147
2015	26,148,672	26,148,672
2016	26,148,672	26,148,672
2017	26,148,672	45,557,694
2018	26,148,672	45,557,694
2019	1,495,850	20,188,873
2020	1,495,850	20,188,873

[127] Memo to Gil Wood, USEPA, from Jill Mozier, EC/R, Inc. Unit Cost Estimates of Residential Wood Heating Appliances. February 21, 2013.
[128] HPBA Solid Fuel Product List. Attachment to E-mail from John Crouch, HBPS, to Gil Wood, EPA. September 24, 2010.

2021	1,495,850	20,188,873
2022	1,495,850	20,188,873

Finally, Tables A-6 and A-7 provide annual costs and emissions, and emission reductions associated with the Proposal and Alternative options, respectively, for each year included in the analyses, including impacts of the rule beyond 2022, and cumulative impacts for each option.

Table A-6. Cost Effectiveness (CE) based on annual and cumulative PM$_{2.5}$ emissions from Central Heaters (Forced Air Furnaces and Hydronic Heating Systems) and Room Heaters (Wood Stoves, Pellet Stoves, and Single Burn Rate Stoves) for the Proposal Option

Year	Annual Capital Costs	Nationwide Annual Cost[1]	Nationwide Average Annual Cost	Annual Snapshots				Emission Reduction, cumulative per year			CE based on total cost & cumulative emission reduction over 20-year stove lifespan (per ton)
				Baseline PM$_{2.5}$ Emissions[2] (tons)	NSPS PM$_{2.5}$ Emissions[2] (tons)	Emission Reduction (tons)	CE based on nationwide average annual cost (per ton)	Baseline PM$_{2.5}$ Emissions (tons)	NSPS PM$_{2.5}$ Emissions (tons)	Emission Reduction (tons)	
2013[3]	$4,754,295	$10,883,300	$3,289,936								
2014[4]	$4,754,295	$10,883,300	$3,289,936	5,587	1,760	3,827	$860	5,587	1,760	3,827	
2015	$4,068,766	$9,314,021	$3,289,936	5,699	1,795	3,903	$843	11,285	3,556	7,730	
2016	$4,068,766	$9,314,021	$3,289,936	5,812	1,831	3,981	$826	17,098	5,387	11,711	
2017	$4,068,766	$9,314,021	$3,289,936	5,929	1,868	4,061	$810	23,026	7,255	15,771	
2018	$4,068,766	$9,314,021	$3,289,936	6,047	1,905	4,142	$794	29,074	9,161	19,913	
2019[4]	$0	$1,325,783	$3,289,936	6,168	464	5,705	$577	35,242	9,624	25,618	
2020	$0	$1,325,783	$3,289,936	6,292	473	5,819	$565	41,534	10,097	31,437	
2021	$0	$1,325,783	$3,289,936	6,417	482	5,935	$554	47,951	10,579	37,372	
2022	$0	$1,325,783	$3,289,936	6,546	492	6,054	$543	54,497	11,071	43,425	
2023	$0	$1,325,783	$3,289,936	6,677	502	6,175	$533	61,174	11,573	49,600	
2024	$0	$1,325,783	$3,289,936	6,810	512	6,298	$522	67,984	12,085	55,899	
2025	$0	$1,325,783	$3,289,936	6,946	522	6,424	$512	74,930	12,607	62,323	
2026	$0	$1,325,783	$3,289,936	7,085	533	6,553	$502	82,016	13,140	68,876	

Year							9.1.8	9.1.9	9.1.10	9.1.11
2027	$0	$1,325,783	$3,289,936	7,227	543	6,684	$492	89,243	13,683	75,560
2028	$0	$1,325,783	$3,289,936	7,372	554	6,818	$483	96,614	14,237	82,377
2029	$0	$1,325,783	$3,289,936	7,519	565	6,954	$473	104,133	14,802	89,331
2030	$0	$1,325,783	$3,289,936	7,669	576	7,093	$464	119,626	15,967	103,659
2031	$0	$1,325,783	$3,289,936	7,823	588	7,235	$455	127,605	16,566	111,039
2032	$0	$1,325,783	$3,289,936	7,979	600	7,380	$446	135,744	17,178	118,566
2033	$0	$1,325,783	$3,289,936	8,139	612	7,527	$437	138,459	16,042	122,417
2034	$0	$1,325,783	$3,289,936	8,302	624	7,678	$429	141,228	14,883	126,345
2035	$0	$1,325,783	$3,289,936	8,468	636	7,831	$420	144,052	13,701	130,352
2036	$0	$1,325,783	$3,289,936	8,637	649	7,988	$412	146,933	12,495	134,439
2037	$0	$1,325,783	$3,289,936	8,810	662	8,148	$404	149,872	11,265	138,607
2038	$0	$1,325,783	$3,289,936	8,986	675	8,311	$396	143,704	10,801	132,903
2039								137,412	10,328	127,084
2040								130,995	9,846	121,149
2041								124,449	9,354	115,095
2042								117,772	8,852	108,920
2043								110,962	8,340	102,622
2044								104,016	7,818	96,198
2045								96,930	7,286	89,645
2046										

Year			
2047	89,703	6,742	82,961
2048	82,332	6,188	76,143
2049	74,812	5,623	69,189
2050	67,143	5,047	62,096
2051	59,320	4,459	54,862
2052	51,341	3,859	47,482
2053	43,202	3,247	39,955
2054	34,900	2,623	32,277
2056	26,433	1,987	24,446
2057	17,796	1,338	16,458
	8,986	675	8,311

	Cumulative Emission Reduction over 20-year stove lifespan	
Nationwide cumulative cost[5]	$85,538,348	1,546,402	$55

[1] Estimated nationwide annual costs are in 2010 $ and are based on a 6-year amortization period of R&D costs at a 7% interest rate (during 2013-2018), plus annual certification and reporting & recordkeeping costs (ongoing through 2038). Years 2039 through 2057 are past the 20-year model design lifespan used in this analysis.

[2] Estimated annual emissions are based on a forecasted revenue growth rate (as a surrogate for shipments) of 2.0 % from 2011 through 2038, in keeping with the average annual growth in real GDP predicted by the US Bureau of Economic Analysis (2013 Global Outlook projections prepared by the Conference Board in November 2012; see http://www.conference-board.org/data/globaloutlook.cfm).

[3] 2013 costs assume manufacturers will begin R&D phase and begin certifying models in anticipation of the 2014 rule compliance date. Emissions in 2013, however, are not included in this analysis because it is prior to the rule compliance date.

[4] Estimated emissions assume Step 1 standard becomes applicable in 2014 and Step 2 standard in 2019. No emission reductions are estimated to result from woodstoves and pellet stoves until 2019, although emission reductions are estimated for all other devices starting in 2014 .

[5] The nationwide cumulative cost represents the cost to manufacturers resulting from the R&D re-design to meet the proposed NSPS and the NSPS-caused certification and reporting & recordkeeping costs to bring these stoves to market from 2013 through 2038. These stoves have lifespans of 20 years or more; thus stoves shipped in 2038 will be emitting through 2057.

Table A-7. Cost Effectiveness (CE) based on annual and cumulative PM$_{2.5}$ emissions from Central Heaters (Forced Air Furnaces and Hydronic Heating Systems) and Room Heaters (Wood Stoves, Pellet Stoves, and Single Burn Rate Stoves) for the Alternative Option

Year	Annual Capital Costs	Nationwide Annual Cost[1]	Nationwide Average Annual Cost	Annual Snapshots				Emission Reduction, cumulative per year			CE based on total cost & cumulative emission reduction over 20-year stove lifespan (per ton)
				Baseline PM$_{2.5}$ Emissions[2] (tons)	NSPS PM$_{2.5}$ Emissions[2] (tons)	Emission Reduction (tons)	CE based on nationwide average annual cost (per ton)	Baseline PM$_{2.5}$ Emissions (tons)	NSPS PM$_{2.5}$ Emissions (tons)	Emission Reduction (tons)	
2013 [3]	$10,295,398	$30,200,681	$10,600,322								
2014 [4]	$10,295,398	$30,200,681	$10,600,322	5,587	1,760	3,827	$2,770	5,587	1,760	3,827	
2015	$8,771,756	$25,731,206	$10,600,322	5,699	1,795	3,903	$2,716	11,285	3,556	7,730	

Year										
2016	$8,771,756									
2017 [4]	$15,388,282	$45,140,228	$10,600,322	5,929	906	5,022	$2,111	23,026	6,294	16,733
2018	$15,388,282	$45,140,228	$10,600,322	6,047	925	5,123	$2,069	29,074	7,218	21,856
2019	$6,824,401	$20,018,806	$10,600,322	6,168	943	5,225	$2,029	35,242	8,161	27,081
2020	$6,824,401	$20,018,806	$10,600,322	6,292	962	5,330	$1,989	41,534	9,123	32,411
2021 [4]	$6,824,401	$20,018,806	$10,600,322							
2022 [4]	$6,824,401	$20,018,806	$10,600,322	6,546	492	6,054	$1,751	54,497	10,596	43,901
2023	$0	$1,325,783	$10,600,322	6,677	502	6,175	$1,717	61,174	11,098	50,076
2024	$0	$1,325,783	$10,600,322	6,810	512	6,298	$1,683	67,984	11,610	56,374
2025	$0	$1,325,783	$10,600,322	6,946	522	6,424	$1,650	74,930	12,132	62,798
2026	$0	$1,325,783	$10,600,322	7,085	533	6,553	$1,618	82,016	12,664	69,351
2027	$0	$1,325,783	$10,600,322	7,227	543	6,684	$1,586	89,243	13,208	76,035
2028	$0	$1,325,783	$10,600,322	7,372	554	6,818	$1,555	96,614	13,762	82,853
2029	$0	$1,325,783	$10,600,322	7,519	565	6,954	$1,524	104,133	14,327	89,806
2030	$0	$1,325,783	$10,600,322	7,669	576	7,093	$1,494	111,803	14,903	96,899
2031	$0	$1,325,783	$10,600,322	7,823	588	7,235	$1,465	119,626	15,491	104,134
2032	$0	$1,325,783	$10,600,322	7,979						

Year										
2033	$0	$1,325,783	$10,600,322	8,139	612	7,527	$1,408	135,744	16,703	119,041
2034	$0	$1,325,783	$10,600,322	8,302	624	7,678	$1,381	138,459	15,567	122,892
2035	$0	$1,325,783	$10,600,322	8,468	636	7,831	$1,354	141,228	14,408	126,820
2036	$0	$1,325,783	$10,600,322	8,637	649	7,988	$1,327	144,052	13,225	130,827
2037	$0	$1,325,783	$10,600,322	8,810	662	8,148	$1,301	146,933	12,981	133,952
2038	$0	$1,325,783	$10,600,322	8,986	675	8,311	$1,276	149,872	12,732	137,140
2039	$0	$1,325,783	$10,600,322	9,166	689	8,477	$1,251	152,870	12,478	140,392
2040	$0	$1,325,783	$10,600,322	9,349	703	8,646	$1,226	155,927	12,219	143,708
2041	$0	$1,325,783	$10,600,322	9,536	717	8,819	$1,202	159,045	11,954	147,091
2042								152,500	11,462	141,037
2043								145,823	10,960	134,863
2044								139,013	10,449	128,564
2045								132,066	9,926	122,140
2046								124,981	9,394	115,587
2047								117,754	8,851	108,903
2048								110,382	8,297	102,086
2049								102,863	7,731	95,132
2050								95,194	7,155	88,039
2051								87,371	6,567	80,804
2053								79,392	5,967	65,897
2054								71,253	5,356	58,220
2055								62,951	4,732	50,388
								54,483	4,095	

Year				
2056		45,846	3,446	42,400
2057		37,037	2,784	34,253
2058		28,051	2,108	25,942
2059		18,885	1,419	17,466
2060		9,536	717	8,819
Nationwide cumulative cost[5]	$307,409,335	Cumulative Emission Reduction over 20-year stove lifespan	1,641,055	$187

[1] Estimated nationwide annual costs are in 2010 $ and are based on 6-year amortization periods of R&D costs at a 7% interest rate (during 2013-2018 for round one R&D and 2017-2022 for round two R&D), plus annual certification and reporting & recordkeeping costs (ongoing through 2041). Years 2042 through 2060 are past the 20-year model design lifespan used in this analysis.

[2] Estimated annual emissions are based on a forecasted revenue growth rate (as a surrogate for shipments) of 2.0 % from 2011 through 2041, in keeping with the average annual growth in real GDP predicted by the US Bureau of Economic Analysis (2013 Global Outlook projections prepared by the Conference Board in November 2012; see http://www.conference-board.org/data/globaloutlook.cfm).

[3] 2013 costs assume manufacturers will begin R&D phase and begin certifying models in anticipation of the 2014 rule compliance date. Emissions in 2013, however, are not included in this analysis because it is prior to the rule compliance date.

[4] Estimated emissions assume Step 1 standard becomes applicable in 2014, Step 2 standard in 2017, and Step 3 standard in 2022. No emission reductions are estimated for woodstoves and pellet stoves until 2017, although emission reductions are estimated for all other devices starting in 2014.

[5] The nationwide cumulative cost represents the cost to manufacturers resulting from the R&D round(s) to meet the proposed NSPS and the NSPS-caused certification and reporting & recordkeeping costs to bring these stoves to market from 2013 through 2041. These stoves have lifespans of 20 years or more; thus stoves shipped in 2041 will be emitting through 2060.

Attachment A

Table AA-1. Non-Catalytic, Catalytic, and Pellet Stove Emissions (HPBA Data)

Rank	Type Cat. / Non. Cat or Pellet	EPA M28 Weighted Avg. Emissions g/hour
1	Non-Cat	0.70
2	Non-Cat	0.80
3	Cat.	0.80
4	Pellet	1.00
5	Cat.	1.10
6	Pellet	1.10
7	Pellet	1.10
8	Non-Cat	1.10
9	Pellet	1.15
10	Pellet	1.20
11	Pellet	1.20
12	Pellet	1.30
13	Non-Cat	1.30
14	Cat.	1.30
15	Cat.	1.35

16	Non-Cat	1.40
17	Cat.	1.40
18	Pellet	1.40
19	Pellet	1.40
20	Non-Cat	1.50
21	Pellet	1.50
22	Non-Cat	1.60
23	Pellet	1.60
24	Pellet	1.60
25	Pellet	1.60
26	Cat.	1.60
27	Pellet	1.67
28	Pellet	1.70
29	Pellet	1.80
30	Non-Cat	1.89
31	Non-Cat	1.90
32	Cat.	1.90
33	Non-Cat	2.00
34	Non-Cat	2.00
35	Pellet	2.00
36	Cat.	2.00
37	Non-Cat	2.00
38	Cat.	2.10
39	Non-Cat	2.10
40	Cat.	2.10
41	Non-Cat	2.10
42	Pellet	2.20
43	Non-Cat	2.30
44	Pellet	2.30
45	Cat.	2.40
46	Cat.	2.40
47	Non-Cat	2.40

48	Non-Cat	2.40
49	Non-Cat	2.40
50	Non-Cat	2.43
51	Non-Cat	2.50
52	Cat.	2.50
53	Non-Cat	2.60
54	Non-Cat	2.60
55	Non-Cat	2.60
56	Pellet	2.60
57	Non-Cat	2.70
58	Non-Cat	2.70
59	Non-Cat	2.70
60	Non-Cat	2.70
61	Non-Cat	2.88
62	Non-Cat	2.90
63	Non-Cat	2.90
64	Non-Cat	2.90
65	Non-Cat	2.90
66	Non-Cat	3.00
67	Non-Cat	3.00
68	Non-Cat	3.00
69	Non-Cat	3.00
70	Non-Cat	3.01
71	Non-Cat	3.06
72	Non-Cat	3.10
73	Pellet	3.10
74	Non-Cat	3.10
75	Pellet	3.10
76	Non-Cat	3.10
77	Non-Cat	3.10
78	Non-Cat	3.10
79	Non-Cat	3.20

80	Non-Cat	3.30
81	Non-Cat	3.35
82	Non-Cat	3.40
83	Non-Cat	3.40
84	Non-Cat	3.40
85	Non-Cat	3.40
86	Non-Cat	3.47
87	Non-Cat	3.50
88	Non-Cat	3.50
89	Non-Cat	3.50
90	Non-Cat	3.50
91	Non-Cat	3.60
92	Non-Cat	3.60
93	Non-Cat	3.60
94	Non-Cat	3.60
95	Non-Cat	3.60
96	Non-Cat	3.60
97	Non-Cat	3.60
98	Non-Cat	3.70
99	Cat.	3.70
100	Non-Cat	3.71
101	Non-Cat	3.72
102	Non-Cat	3.80
103	Non-Cat	3.80
104	Non-Cat	3.80
105	Non-Cat	3.90
106	Non-Cat	3.90
107	Non-Cat	4.00
108	Non-Cat	4.00
109	Non-Cat	4.00
110	Pellet	4.00
111	Non-Cat	4.10

112	Cat.	4.10
113	Non-Cat	4.10
114	Non-Cat	4.10
115	Non-Cat	4.18
116	Non-Cat	4.19
117	Non-Cat	4.20
118	Non-Cat	4.20
119	Non-Cat	4.20
120	Non-Cat	4.30
121	Non-Cat	4.30
122	Non-Cat	4.31
123	Non-Cat	4.40
124	Non-Cat	4.40
125	Non-Cat	4.40
126	Non-Cat	4.40
127	Non-Cat	4.40
128	Non-Cat	4.50
129	Non-Cat	4.50
130	Non-Cat	4.50
131	Non-Cat	4.70
132	Non-Cat	4.80
133	Non-Cat	4.80
134	Non-Cat	5.10
135	Non-Cat	5.20
136	Non-Cat	5.30
137	Non-Cat	5.50
138	Pellet	5.50
139	Non-Cat	5.90
140	Non-Cat	6.00
141	Non-Cat	6.00
142	Non-Cat	6.10
143	Non-Cat	6.90

| 144 | Non-Cat | 7.30 |
| 145 | Non-Cat | 7.50 |

Table AA-2. Catalytic Stove Emissions (HPBA Data)

Rank	Type Cat. / Non. Cat or Pellet	EPA M28 Weighted Avg. Emissions g/hour
1	Cat.	0.80
2	Cat.	1.10
3	Cat.	1.30
4	Cat.	1.35
5	Cat.	1.40
6	Cat.	1.60
7	Cat.	1.90
8	Cat.	2.00
9	Cat.	2.10
10	Cat.	2.10
11	Cat.	2.40
12	Cat.	2.40
13	Cat.	2.50
14	Cat.	3.70
15	Cat.	4.10

Table AA-3. Pellet Stove Emissions (HPBA Data)

Rank	Type Cat. / Non. Cat or Pellet	EPA M28 Weighted Avg. Emissions g/hour
1	Pellet	1.00
2	Pellet	1.10
3	Pellet	1.10
4	Pellet	1.15
5	Pellet	1.20
6	Pellet	1.20
7	Pellet	1.30
8	Pellet	1.40
9	Pellet	1.40
10	Pellet	1.50
11	Pellet	1.60
12	Pellet	1.60
13	Pellet	1.60
14	Pellet	1.67
15	Pellet	1.70
16	Pellet	1.80
17	Pellet	2.00
18	Pellet	2.20
19	Pellet	2.30
20	Pellet	2.60
21	Pellet	3.10
22	Pellet	3.10
23	Pellet	4.00

24 | Pellet | 5.50